C++ PROGRAMMING WITH APPLICATIONS IN

Administration, Finance and Statistics

C++ PROGRAMMING WITH APPLICATIONS IN
Administration, Finance and Statistics

Willi-Hans Steeb & Fritz Solms
Rand Afrikaans University, Johannesburg, South Africa

World Scientific
Singapore • New Jersey • London • Hong Kong

Published by

World Scientific Publishing Co. Pte. Ltd.

P O Box 128, Farrer Road, Singapore 912805

USA office: Suite 1B, 1060 Main Street, River Edge, NJ 07661

UK office: 57 Shelton Street, Covent Garden, London WC2H 9HE

British Library Cataloguing-in-Publication Data
A catalogue record for this book is available from the British Library.

C++ PROGRAMMING WITH APPLICATIONS IN ADMINISTRATION, FINANCE AND STATISTICS

ISBN 981-02-4066-X

This book is printed on acid and chlorine free paper.

Printed in Singapore by FuIsland Offset Printing

Preface

C++ is the most widely available and used object-oriented programming language. It is a very useful language that is successfully utilized by many programmers in many application areas. It is a reasonably carefully thought-out language where the design is based partly on acknowledged principles and partly on solid experience and feedback from actual use. The language is much more complicated and complex than, say, C and PASCAL. The key concepts of data abstraction and object-oriented programming are new to most people. One can use C++ effectively as a more strongly type checked C with a small amount of simple data abstraction and library use. This programming language has all the tools to write compact and comprehensive programs for finance, administration and statistics. Many of the most experienced individuals and organizations in the industry and academia use C++. Many programmers strongly prefer C++ over procedural languages such as C and PASCAL and deem the feature set provided by C++ as close to essential for their applications. C++ supports object-oriented programming. It supports it pretty well for real life applications. Its type system and general model of the world are coherent and relevant to real applications. C++ and its various implementations have a solid core where different compilers tend to agree and where the code generated is reliably good.

This book provides software solutions to problems in finance, administration and statistics. Most standard methods used in these fields are given together with their programs in C++. In some cases the output of the programs is also displayed. Moreover the book gives an introduction to C++. Most of the programs are written in C++. Some of the programs are written in C. All the standard methods such as number sorting, name sorting, applications of function templates, applications of class templates, inheritance, virtual classes, string manipulations are included.

In chapter 2 we give some basic programs in C++. These programs are very helpful for the beginners. They are also the building blocks for more complex applications. Chapter 3 is devoted to the manipulation of strings. The class concept is introduced in chapter 4. Chapter 5 and 6 deal with function and class templates. A collection of introductory programs are given. Sorting and searching algorithms are implemented in C++ in chapter 7. Chapter 8 gives a collection of useful classes. They include string classes, vector classes, a matrix class, linked lists, trees and a date class. Chapter 9 is devoted to output/input streams. Here we consider C and C++ programs. In chapter 10 a collection of applications in finance are given. Applications in administration are considered in chapter 11. Problems in statistics are considered in chapter 12. Chapter 13 describes exception handling. The standard template library is introduced in chapter 14 and several examples are given.

The level of presentation is such that one can study the subject early on in ones education in science. There is a balance between practical programming and the underlying language. The book is ideally suited for use in lectures on C, C++ and object-oriented programming. The beginner will also benefit from the book. Almost all C++ programs in the book comply to the ANSI C++ standard.

The reference list gives a collection of text books useful in the study of the computer language C++. There are a number of good text books for C++ available [3], [6], [7], [8], [11], [13], [9]. We also refer to *The Annotated C++ Reference Manual* by Margret Ellis and Bjarne Stroustrup [6]. For data structures we refer to Budd [4]. For applications in science we refer to Tan Kiat Shi and W.-H. Steeb [14] and Steeb [12]. The standard template library is described by Schildt [10], Ammeraal [1] and Budd [5].

For a comprehensive discussion of the Boyer-Moore algorithm, the Floyd-Warshall algorithm, Dijkstra algorithm, the sorting methods and the divide-and-conquer strategy we refer to Baase (1989) [2].

Without doubt, this book can be extended. If you have comments or suggestions, we would be pleased to have them. The email address of the first author is:

`whs@na.rau.ac.za`

The web page of the first author is:

`http://zeus.rau.ac.za/`

One of the authors (W.-H. Steeb) thanks his students from the Computational Science Programme of the National University of Singapore for stimulating discussions on object-oriented programming. They are: Fong Wei Leng, Heng Kok Liang, Tan Kiat Shi, Lee Wan Sie, Gan Chee Kwan, Toh Che Chou, Koh Mei Leng, Lee Yih, And Mieng Hwang, Teo Wee Chung, Lee Choon Meng.

Keywords in C++

C++ has the following keywords:

Table: C++ language keywords

asm	double	new	switch
auto	else	operator	template
break	enum	private	this
case	extern	protected	throw
catch	float	public	try
char	for	register	typedef
class	friend	return	union
const	goto	short	unsigned
continue	if	signed	virtual
default	inline	sizeof	void
delete	int	static	volatile
do	long	struct	while

Remark. `asm` is reserved for implementors to include a machine specific facility to pass information to an assembler. `extern` indicates that external linkage is to be made to this item. Obviously the placement of such an item must be known at compile time and hence must be implicitly static.

Basic Data Types of C++

The basic data types or primitive data types are as follows:

Type	May also be specified as	Comment
char		Holds a character
unsigned char		Unsigned version of char
signed char		Signed version of char
int	signed int signed	A machine word that holds an integer number
unsigned int	unsigned	Unsigned version of int
short	signed short int signed short	May have less precision than an int
unsigned short	unsigned short int	Unsigned version of short
long	long int	May have more precision than an int
unsigned long	unsigned long int	Unsigned version of long
long int	signed long int	
float		A number held in floating point form
double		May have more precision than a float
long double		May have more precision than a double

Note: A float, double and long double are stored as a floating point number. The rest of the basic data types are stored as an exact quantity.

The unsigned types will be able to store a larger positive number than their signed types.

The precision of items declared with these data types may vary, depending on the compiler or machine used.

The representation of char may be either signed or unsigned. This is to allow implementors the choice of the architectural representation of a char.

As can be seen from the table above, there is no basic data type for boolean values. C++ like C before, uses integer values to represent true and false. Newer C++ compilers have a boolean data type called bool.

Truth value	C++ delivers	C++ takes a truth value
true	1	Any value other than 0
false	0	0

C++ Operators

The following *precedence table* shows a summary of C++ operators in order of decreasing precedence. Each box holds operators of the same level of precedence.

For example

```
int u, x, y, z;
x = 3;  y = 9;  z = 2;
u = x + y % z;
```

firstly evaluates the remainder of `y%z` and then adds this to `x` and assigns the result to `u`. Another example is

```
*g.value          (*g).value
```

The member selection operator `.` has higher precedence then the dereferencing operator `*`. Hence `*g.value` dereferences `g.value`, whereas `(*g).value` dereferences `g` and then selects the member value of the object to which `g` points. The expression `(*g).value` can also be written as `g -> value`. The expression

```
int r = m%n/p;
```

first evaluates `m%n` then we divide and finally the assignment is done. Thus the right hand side evaluates from left to right.

The use of the operators given in following table will be illustrated with sample programs throughout this text. In C++ one can define operators for ones own data types. For example if one defines a data type for rational numbers, then one can define the addition, subtraction, multiplication etc. for this data type. This facility is called operator overloading. All operators can be overloaded (see discussion on operator overloading later in the text) except for

```
    .    .*    ::    ?:
```

When overloading operators, one should bear in mind that the order of precedence and the syntax remains the same as that for the basic data types.

Operator	operator name	example
::	scope resolution	*class_name*::*member*
::	global	::*name*
++	preincrement	++*lvalue*
--	predecrement	-- *lvalue*
.	member selection	*object*.*name*
->	member selection	*pointer*->*member*
[]	array subscription	*pointer*[*expr*]
()	function call	*expr*(*expr_list*)
()	value construction	*type*(*expr_list*)
sizeof	size of object	sizeof *expr*
sizeof	size of type	sizeof(*type*}
~	one's complement	~*expr*
!	not	!*expr*
+ , -	unary plus and minus	-*expr* and +*expr*
&	address of	&*lvalue*
*	dereferencing	**expr*
new	create (allocate memory)	new *type*
delete	deallocate pointer memory	delete *pointer*
delete[]	free array memory	delete[] *pointer*
()	cast (type conversion)	(*type*) *expr*
.*	member section	*object*.**pointer-to-member*
->*	member section	*pointer*->**pointer-to-member*
* , / , %	multiply, divide, modulo (remainder)	*expr* / *expr*
+ , -	add, subtract	*expr* + *expr*
<< , >>	shift left , shift right	*expr* << *expr*
< , <= , > , >=	relational operators	*expr* < *expr*
==	equal	*expr* == *expr*
!=	not equal	*expr* != *expr*
&	bitwise AND	*expr* & *expr*
^	bitwise exclusive OR	*expr* ^ *expr*
|	bitwise inclusive OR	*expr* | *expr*
&&	logical AND	*expr* && *expr*
||	logical inclusive OR	*expr* || *expr*
? :	conditional expression	*expr* ? *expr* : *expr*
=	simple assignment	*lvalue* = *expr*
*= , /= , += , -=	multiply and assignment, ...	*lvalue* *= *expr*
<<= , >>=	shift left (right) and assign	*lvalue* <<= *expr*
&= , |= , ^=	AND (OR, XOR) and assignment	*lvalue* &= *expr*
throw	throw exception	throw *expr*
,	comma (sequencing)	*expr* , *expr*
++	postfix increment	*lvalue*++
--	postfix decrement	*lvalue*--

Contents

1 Introduction 1

1.1 Aim of Object-Oriented Programming 1

1.1.1 Information Hiding 2

1.1.2 Inheritance 2

1.1.3 Polymorphism 3

1.1.4 Templates 3

1.1.5 Standard Template Library 3

1.2 Why C++ ? 4

2 C++ Basics 5

2.1 Introduction 5

2.2 A Simple C++ Program 7

2.3 Basic Data Types 9

2.4 Arithmetic Operations 11

2.5 ASCII Table and Types Conversion 14

2.6 Precedence Table 16

2.7 Pointers and References 18

2.8 Control Statements 21

2.8.1 Introduction 21

2.8.2 The if Statement 22

2.8.3 The for Loop, while Loop, do-while Loop 24

2.8.4 The switch Statement 26

2.9 Arrays 28

2.10 Strings 33

2.11 The sizeof Operator 37

2.12 The new and delete Operators 38

2.13 Arrays of Pointers 42

2.14 Logical AND, Logical OR and Logical NOT 46

2.15 Pass by Value, Pass by Reference 48

2.16 Command-Line Arguments 51

2.17 Bitwise Operations 55

2.18 Shift Operations 57

2.19 Passing a Function to a Function 59

2.20 Recursion 60

2.21 Inline Assembly Language 63

2.22 Time Header File 64
2.23 Random Numbers 66
2.24 Constants and Inline Functions 67
2.25 Jump Statements 69
2.26 The static Keyword 71
2.27 Structures 72

3 String Manipulations **75**
3.1 Introduction 75
3.2 Digit Conversion to Character and String 76
3.3 Character in String 77
3.4 Append a Character to a String 78
3.5 Searching for a Character in a Given String 79
3.6 Delete a Character from a String 80
3.7 Replacement of a Character by a Character 81
3.8 Swapping Characters in a String 82
3.9 First n Characters from a String 83
3.10 Substring in String 84
3.11 Assignment and the Function strcpy 85
3.12 Permutations of a Word 86
3.13 String within String 88
3.14 Comparing Strings 90
3.15 Length of a String 92
3.16 Permutations of Characters in a String 93
3.17 Swapping Parts of a String 95
3.18 String Handling using Recursion 96
3.19 Concatenating Strings 98
3.20 Replacing a Substring in a String 100
3.21 Reversing the Characters in a String 102
3.22 Boyer and Moore's Algorithm 104
3.23 Subsets of a Given Set of Strings 107
3.24 Counting Characters and Digits in a Phrase 108
3.25 Counting Characters and Words in a Phrase 109
3.26 Centering Text 111
3.27 The Function getline 112

4 The Class Concept **113**
4.1 Introduction 113
4.2 What is a Class ? 115
4.3 Constructors and Destructors 116
4.4 Copy Constructor 119
4.5 Operator Overloading 120
4.6 The Keyword this 121
4.7 Examples 123
4.7.1 Example 1. 123

4.7.2 Example 2. 126
4.7.3 Example 3. 128
4.7.4 Example 4. 130
4.7.5 Example 5. 133
4.8 Friend 135
4.9 Derivation and Inheritance 137
4.10 Virtual Functions 140

5 Function Templates **143**
5.1 Introduction 143
5.2 Examples 144

6 Class Templates **151**
6.1 Introduction 151
6.2 Examples 152
6.2.1 A Template Vector Class 152
6.2.2 A Template Stack Class 156
6.2.3 A Template Stack Class with Node 158
6.2.4 Two Generic Data Types 162

7 Sorting and Searching **163**
7.1 Introduction 163
7.2 Largest Number in an Array 164
7.3 Quicksort and Bubblesort 165
7.4 Shellsort 167
7.5 Using the Function bsearch 171
7.6 Using the Function lsearch 172
7.7 Name Sorting 173
7.8 Name and Number Sorting Using Templates 177
7.9 Student Names and Student Numbers 181
7.10 Search for Names in a List 183
7.11 Names of Employees in a Table 184
7.12 The Divide-and-Conquer Methodology 186
7.13 Sorting Vectors 187

8 Useful Classes **189**
8.1 Introduction 189
8.2 String Class 190
8.2.1 Introduction 190
8.2.2 A Simple String Class 191
8.2.3 Advanced String Classes 194
8.2.4 Standard Library and String Class 200
8.3 Vector Class 204
8.3.1 Introduction 204
8.3.2 A Simple Vector Class 205

8.3.3 Construction of an Advanced Vector Class . . . 209
8.3.4 Listing of Vector Class . . . 217
8.4 Matrix Class . . . 225
8.4.1 Introduction . . . 225
8.4.2 Listing of Matrix Class . . . 228
8.5 Linked Lists . . . 237
8.5.1 Introduction . . . 237
8.5.2 Simple Linked Lists . . . 241
8.5.3 Singly Linked Circular List . . . 243
8.5.4 Recursive List . . . 245
8.5.5 Linked List and Templates . . . 248
8.5.6 Linked List and Iterators . . . 250
8.5.7 Intrusive List . . . 258
8.5.8 Generalized Lists . . . 263
8.6 Binary Trees . . . 269
8.6.1 Introduction . . . 269
8.6.2 A Simple Binary Tree . . . 270
8.6.3 Binary Trees and Templates . . . 272
8.7 A Set Class . . . 276
8.8 A Date and Month Class . . . 282
8.8.1 Introduction . . . 282
8.8.2 Listing of Complete Month and Date Classes . . . 296

9 File Manipulations 305

9.1 Introduction . . . 305
9.2 C Style File Manipulations . . . 308
9.2.1 Output to File . . . 308
9.2.2 Reading String from File . . . 309
9.2.3 Word Count in File . . . 310
9.2.4 File Searching Program . . . 311
9.2.5 Counting Characters in a File . . . 313
9.2.6 Class and Standard Output . . . 316
9.2.7 The Functions fseek, fread, fwrite . . . 318
9.2.8 The Function fscanf . . . 321
9.3 C++ Style File Manipulation . . . 323
9.3.1 Output to File . . . 323
9.3.2 Reading String from File . . . 324
9.3.3 String Replacement in File . . . 325
9.3.4 Reading and Writing of File . . . 327
9.3.5 Writing File to Another File . . . 331
9.4 Binary File Operations . . . 332
9.5 The Functions seekp and seekg . . . 334
9.6 The Function isatty . . . 335
9.7 The Function getenv . . . 336

10 Applications in Finance **337**
10.1 Introduction . . . 337
10.2 Payroll . . . 338
10.3 Loan and Interest . . . 340
10.4 Amortization . . . 343
10.5 Calculating the Loan Period . . . 346
10.6 Periodic Savings . . . 348
10.7 Compound Amount . . . 352
10.8 The Floyd–Warshall Algorithm . . . 356
10.9 The Dijkstra Algorithm . . . 359
10.10 Linear Programming and Simplex Method . . . 362

11 Applications in Administration **375**
11.1 Introduction . . . 375
11.2 Names, Address and Inheritance . . . 376
11.3 Line Editor and Nested Classes . . . 379
11.4 Mergesort Applied to a List . . . 386
11.5 Linked List of Strings . . . 390
11.6 Data Encryption . . . 394
11.7 Vector Class and a Dictionary . . . 397
11.8 Spelling Checker . . . 399
11.9 Huffman Code . . . 403
11.10 Eliza . . . 409
11.11 Counting the Letters in a Phrase . . . 414
11.12 The Farmer, Wolf, Goat and Cabbage Problem . . . 420
11.13 Roman Numbers . . . 427
11.14 Day of the Week . . . 430
11.15 Gregorian Calender . . . 431
11.16 Records . . . 436
11.17 Finite State Machine . . . 443
11.18 Realtor . . . 446

12 Applications in Statistics **453**
12.1 Introduction . . . 453
12.2 Arithmetic, Harmonic and Geometric Mean . . . 454
12.3 Mean Value and the Standard Deviation . . . 458
12.4 Histogram for Series of Numbers . . . 461
12.5 Data Fitting by Least Squares . . . 463
12.6 Curve Fitting . . . 467
12.7 Pseudo-Random Number Generator . . . 473
12.8 Correlation Coefficient . . . 475
12.9 Autocorrelation Function . . . 477
12.10 Random Number Generator and π . . . 480

13 Exception Handling **485**
13.1 Introduction 485
13.2 Error States 485
13.3 Exception Handling 486

14 The Standard Template Library **503**
14.1 Introduction 503
14.2 The Namespace Concept 505
14.3 The Vector Class 506
14.4 The List Class 510
14.5 The Stack Class 512
14.6 The Queue Class 514
14.7 The Deque Class 516
14.8 The Bit Set Class 518
14.9 The Set Class 520
14.10 The Map Class 522
14.11 The Algorithm Class 525

Bibliography **527**

Index **529**

Chapter 1

Introduction

1.1 Aim of Object-Oriented Programming

Object-oriented programming is the most dramatic innovation in software development in the last decade. It offers a new and powerful way to cope with the complexity of programs. Using object-oriented programming we can write clearer, more reliable, more easily maintained programs. Integrating data and functions (methods) is the central idea of object-oriented programming.

Object-oriented programming methods aim to achieve the following:

- To simplify the design and implementation of complex programs.
- To make it easier for teams of designers and programmers to work on a single software project.
- To enable a high degree of reusability of designs and of software codes.
- To decrease the cost of software maintenance.

In order to achieve these aims, object-oriented programming languages are expected to support a number of features:

1. Information hiding (encapsulation)
2. Polymorphism
3. Inheritance

Each of these concepts is discussed below.

1.1.1 Information Hiding

Information hiding is achieved by restricting the access of the user to the underlying data structures to the predefined methods for that class. This shifts the responsibility of handling the data fields correctly from the user of the code to the supplier. For example, a programmer using a Date class is usually not interested in the implementation details. These include the underlying data structure of Date (i.e. whether the month is stored as an integer, a string or an enumeration type) and the underlying code (i.e. how two dates are subtracted from each other). If, at a later stage a more efficient way of storing the date or of calculating the number of days between two dates is introduced, this should not affect the programmers who have been using the date class. They should not have to search through all their programs for any occurrence of Date and to make the relevant changes. Such a maintenance nightmare and the very high cost accompanying it is prevented by information hiding where access to the underlying data structures is given only via the predefined methods of the class. A further advantage of information hiding is that it can be used to guarantee the integrity of the data (e.g. to prevent the user from setting the month equal to 13). C++ and Java provide the keyword `private` and `public` for data hiding.

1.1.2 Inheritance

Often one object shares a number of characteristics with another, but has a few additional attributes. For example, your firm might have a data base for vehicles. For each vehicle the registration number, model, year and maintenance history are stored. Suppose the firm wants to expand this data base to include its trucks. However, they want to store additional information for trucks, such as the the payload and the number of axes (for toll-road purposes). Instead of rewriting all the code for trucks, one can see a truck as a special case of a vehicle (having all the attributes of a vehicle plus a few additional attributes). By making truck a derived class of vehicle, it inherits all the attributes of vehicle and the programmer has only to add the code for the additional attributes. This not only reduces the amount of coding to be done and maintained, but it also ensures a higher level of consistency. Should the country decide to change its format for number plates, it has to be only changed in the base class vehicle and the trucks automatically inherit the change.

1.1.3 Polymorphism

Often one wants to perform an action such as editing on a number of different objects (i.e. text files, pixel files, charts, etc.). Polymorphism allows one to specify an abstract operation like editing, leaving the actual way in which this operation is performed to a later stage. In the case where dynamic binding (linking during run-time) is used the decision of which code to be used for the editing operation is only made during run-time. This is especially useful in the case where the actual type of object which is to be edited is only known at run-time.

Furthermore, if a new object type is to be supported (e.g. faxes) then the new editing code (provided by the fax-software supplier) can be linked in at run-time. Even when the original program is not recompiled, it can support future additions to the system. Polymorphism is hence again a higher level of abstraction, allowing the programmer to specify an abstract action to be performed on abstract objects.

1.1.4 Templates

A less abstract form of polymorphism is provided via template support. In many instances a similar algorithm is needed for many different data structures. For example, sorting algorithms are essentially the same for integers, floating point numbers, names or any other data structure. Similarly, matrix multiplication is essentially the same for integers, floating point numbers and complex numbers. Traditionally one would have to write a sorting or matrix-multiplication algorithm for every data type. Any change in the algorithm must then be implemented in all the copies of that algorithm. Templates allow one to define a single algorithm for many different data types. The actual data types are resolved during compiling and hence there are no run-time overheads.

1.1.5 Standard Template Library

The Standard Template Library (STL) provides many useful and generally applicable programming tools. At the core of the Standard Template Library are three foundational items: containers, algorithms, and iterators. The classes included are: `algorithm`, `deque`, `functional`, `iterator`, `list`, `map`, `memory`, `numeric`, `queue`, `set`, `stack`, `utility` and `vector`.

1.2 Why C++ ?

Firstly, C++ supports the central concepts of object orientated programming: encapsulation, inheritance and polymorphism (including dynamic binding). It has good support for dynamic memory management and supports both, procedural and object-orientated programming.

However, other well designed programming languages have failed against relatively poor competitors. Being a good programming language is not sufficient for survival. An additional important requirement for a powerful programming language is portability. If a firm replaces one computer system with another, only a minimal amount of recording (if any) should be required. C++ compilers are available for virtually all machines and a high level of compatibility is ensured by the ANSI-standard for C++. Furthermore, there exists a huge amount of good C-code. C++ supports C as a subset and hence this huge amount of code can be reused in new software projects.

All these points have contributed to C++ being the fastest growing computer language for nearly all computer and operating systems and for nearly all software applications, ranging from scientific to administrative programs to real-time industrial applications and computer games.

Chapter 2

C++ Basics

2.1 Introduction

In this chapter we survey C++ basics using simple programs. In section 2.2 we give a simple C++ program to explain `cout` and `cin` for output and input. Section 2.3 is devoted to basic data types also called primitive data types. Arithmetic operations are introduced in section 2.4. The ASCII table is used to do type conversion from `char` to `int` and vice versa. Type conversion and the ASCII table are described in section 2.5. A command in C and C++ can involve more than one operator, for example, `=`, `+` and `*`. The precedence table provides the information about the order of the operations. In section 2.6 the use of the precedence table is explained. Section 2.7 gives a detailed discussion of pointers and references. Pointers hold the address of some location in the computer memory, which in turn holds information, for example an integer. In other words, it points to the date, like an address book or an index. A pointer must be declared as pointing to some particular type, even if that type is `void` (which really means a pointer to anything). Control statements and selection statements are introduced in section 2.8. Arrays of basic data types are reviewed in section 2.9. Strings are arrays of characters, where the end of the string is indicated by the null character `'\0'`. Strings are considered in section 2.10. The `sizeof` operator finds the size (in bytes) of a data type or pointer. In section 2.11 we give some applications of the `sizeof` operator. Dynamic memory allocation is studied in section 2.12 with one, two and three dimensional arrays being considered. The `new` and `delete` operators are introduced. The `new` and `delete` operators offer dynamic storage allocation and deallocation. The operator `new` tries to create an object by allocating enough memory in free store (also called the heap). The storage duration of the new object is from the point of creation until the operator `delete` kills it by deallocating its memory, or until the end of the program. Section 2.13 is devoted to arrays of pointers. A multidimensional array can be expressed in terms of an array of pointers rather than as a pointer to a group of contiguous arrays. The logical AND, logical OR and logical NOT is introduced in section 2.14. In using functions in C and C++ the concept of pass by value and pass by reference is of central importance. This concept is discussed in section 2.15 together with pitfalls.

Section 2.16 illustrates the use of command-line parameters. Two parameters can be passed to `main`: `argc` and `argv`. `argc`, an integer, is the number of command-line arguments passed to `main`. `argv` is an array of pointers to strings. Bitwise operations and shift operations are introduced in section 2.17 and 2.18, respectively. The functions `setbit` and `clearbit` is implemented. In section 2.19 we show how to pass a function to a function. The concept of recursion is introduced in section 2.20 and three example are given. Many C and C++ compilers allow inline assembly language. In section 2.21 we give an example. The time header file is introduced in section 2.22 and an application of the methods `ctime` and `difftime` is given. In section 2.23 we show how ANSI C and ANSI C++ handles random numbers using the time header file. Section 2.24 illustrates the use of constants and inline functions. It also explains why macros should be avoided. Jump statements such as `goto` are discussed in section 2.25. The keyword `static` is explained in section 2.26 and an example is given. Structures are collection of different data types. They are introduced in section 2.27 and an example is given. Structures can be considered as a special case of a class.

2.2 A Simple C++ Program

Any C++ program has at least one function, the function `main`. This function is generally referred to as the main program. The body of the function is enclosed within curly brackets {...} and each statement is terminated by a semicolon. The curly brackets perform the function of the `begin` ... `end` block in other languages. Note that after `//` the remainder of the line is treated as a comment. Furthermore, anything between `/*` and `*/` is also treated as a comment. This comment is also used in C.

Consider now the following simple C++ program.

```
// simple.cpp    (file name)

#include <iostream.h> // including library iostream for cout
                      // cout is used to write onto the terminal
                      // cin is used to read from the keyboard

void main(void)     // start of the main program, program starts here
{                   // block begin
int a;              // this declares an integer,
                    // provides memory location
a = 7;              // a is intialized to the integer 7,
int b;              // this declares another integer
b = 4;              // assignment of integer 4
int sum;            // declares another integer which holds
                    // the sum of the two integers a and b
sum = a + b;        // this line calculates the sum of a and b
                    // and puts the result into the
                    // memory location sum

// Now we put the result on the screen using the following command
cout << "sum = " << sum << endl;  // the string "sum" and then sum
                                  // is pushed onto the standard
                                  // output stream (computer screen).

// reading input from keyboard
int x;
cout << "enter an integer: ";
cin >> x;
cout << "The integer entered is: " << x;
}                                 // block end (end program)
```

The `include` compiler directive tells the compiler to include source files or header files in the compilation, thereby enabling the programmer to make use of standard C++-libraries, or libraries developed by himself or anyone else. In the above example output to the screen is done via the standard output stream `cout`. This stream is `cout` defined in the standard C++ library `iostream` whose header file `iostream.h` defines the interface for the library. The data are read in from the standard input stream (usually the keyboard) via `cin` and pushed into the relevant variables via the `>>` operator. A C and C++ program has standard streams associated with it. They are automatically set up by the runtime environment when we include `<stdio.h>`. The three standard streams are:

`stdin` The input device from which the program normally receives its data. For example the library function `getchar` uses `stdin`.

`stdout` The output device to which the program normally directs its output. For example, the library function `printf` uses `stdout`.

`stderr` The output device to which the program directs its diagnostic messages.

When a C++ program uses the I/O Stream class, the following predefined streams are also provided in addition to the standard streams.

`cin` The standard input stream.

`cout` The standard output stream.

`cerr` The standard error stream.

`clog` Also the standard error stream. Output to this stream is fully buffered.

C++ *statements* are expressions terminated by a semicolon. They can be grouped together within curly brackets `{ ... }`. Any set of statements can be grouped within curly brackets.

Variables in C and C++ can be defined anywhere in the code. Generally it is a good idea to define variables as close to the point where they are used as is possible. Global variables should be used only in exceptional cases. The scope of any variable declared within a set of curly brackets is restricted to within these curly brackets.

Functions are defined with a header defining the interface (i.e. what the function receives and what it returns) and a body enclosed in curly brackets. The `main` program is itself a function with the name `main`. In our example the main program receives no parameters or variables and returns nothing (`void`).

2.3 Basic Data Types

Data types are constants, variables, and structures that contain numbers (integer and real), text (characters and strings) or addresses (of variables and structures). In chapter 4 we introduce abstract data types. The *basic data types* in C++ are

```
short, int, long,  float, double, char.
```

C and C++ allow us to declare the basic data types `short`, `int`, `long` and `char` to be unsigned

```
unsigned short, unsigned int, unsigned long, unsigned char
```

This means that instead of having negative values, these types only contain non-negative values (greater than or equal to zero).

The range of the data types is as follows. For `short` we have 2 bytes and the range −32768 to 32767. The size of `int` is implementation dependent. Under DOS we have the size of `int` is 2 bytes. Under Linux, OS/2 and Windows NT it is 4 bytes. C and C++ have a `sizeof` operator which provides the size of data types and pointers. The data types `float` is of size 4 bytes with range `+/-3.4E 38` (7 digits). The data type `double` is of size 8 bytes and has the range `+/-1.7E 308` (15 digits). Some compilers also allow `long double`. The size of `long double` is 10 bytes. The data type `char` is a single *byte*, big enough to hold a single character. Thus the range for `char` is −128 to 127 and for `unsigned char` is 0 to 255.

There is no boolean data type in C. We identify 1 with *true* and 0 with *false*. Newer C++ compiler have a data type `bool`. Variables of any data type can be declared constant:

```
  const double pi = 3.141592;
```

If a certain variable should always remain constant it should be declared so. This enables the compiler to report any violation of the constantness and thus enhances program integrity.

All integer values are displayed in decimal. In *hexadecimal notation* we have the prefix `0x`. In this notation we identify

```
10 -> A, 11 -> B, 12 -> C, 13 -> D, 14 -> E, 15 -> F.
```

Thus the hexadecimal number `0xFF` is 255 in decimal. Thus we can declare and define

```
int i = 0xAF88;
```

Addresses are displayed in hexadecimal notation.

```
// basic.cpp

#include <iostream.h>

int main()
{
  short a = 145;   short b = 234;
  short c = a + b;
  cout << "c = " << c << endl;   // => 379

  int d = 23456;   int e = 5678;
  int f = d + e;
  cout << "f = " << f << endl;   // => 29134

  long i = 567890;   long j = 3456;
  long k = i + j;
  cout << "k = " << k << endl;   // => 571346

  long m = 2147483647;   long n = 1;
  long p = m + n;
  cout << "p = " << p << endl;  // => - 2147483648 why ?

  unsigned long x = 4294967295;
  unsigned long y = 1;
  unsigned long z = x + y;
  cout << "z = " << z << endl; // => 0 why ?

  float r = 3.13;   float s = 4.5;
  float t = r + s;
  cout << "t = " << t << endl; // => 7.63

  double u = -0.0123;
  double v = 1.371;
  double w = u + v;
  cout << "w = " << w << endl; // => 1.3587

  char c1 = 'X';  char c2 = 'Y';
  cout << c1 << c2;  // => XY

  return 0;
}
```

2.4 Arithmetic Operations

Operations assign one value to another, combine values (add, divide, etc.) and compare values (equal, not equal, etc.). The data items that operators act upon are called operands. The operands acted upon by arithmetic operators must represent numeric values. Thus the operands can be integer quantities, floating-point or characters. We recall that character constants represent integer values, as determined by the computer's character set (ASCII table).

The arithmetic operators are

```
+,  -,  *,  /,  %,  ++,  --
```

The *remainder operator* % is sometimes referred to as the *modulus operator*. The modulus operator can, for example, be used to test whether an integer number is odd or even. The remainder operator % requires that both operands be integers and the second operand be nonzero. Similarly, the division / requires that the second operand be nonzero, though the operands need not be integers. Division of one integer quantity by another is referred to as *integer division*. This operation always results in a truncated quotient (i.e. the decimal portion of the quotient will be dropped). On the other hand, if the division operation is carried out with two floating point numbers, or with one floating-point number and one integer, the result will always be a floating point quotient. The increment operator ++ adds 1 to any value. The decrement operator -- subtracts 1 from any value.

C, C++ and Java allows us to use a shorthand when writing expressions that contain multiple operators. The combined operators are

```
+=,  -=,  *=,  /=,  %=
```

Thus the line

```
int a += b;
```

is equivalent to

```
int a = a + b;
```

```
// arith.cpp

#include <iostream.h>

int main()
{
   // addition +
   int a = 7; int b = -9;
   int c = a + b;
   cout << "c = " << c << endl; // => -2

   // combined operator
   int d = 123;
   d += 500; // equivalent to d = d + 500;
   cout << "d = " << d << endl; // => 623

   // multiplication *
   unsigned long e = 23;
   unsigned long f = 45;
   unsigned long g = e*f;
   cout << "g = " << g << endl; // => 1035

   unsigned long h = 11;
   h *= 20; // equivalent to h = h*20;
   cout << "h = " << h << endl; // => 220

   // integer division
   int i = 17;
   int j = 3;
   int k = i/j; // integer division
   cout << "k = " << k << endl; // => 5

   int m = 17;
   m /= 7; // equivalent to m = m/7;
   cout << "m = " << m << endl; // => 2

   // modulus (remainder) %
   int n = 17%7;
   cout << "n = " << n << endl; // => 3

   int p = 56;
   p %= 6; // equivalent to p = p%6;
   cout << "p = " << p << endl; // => 2
```

```
    // increment and decrement by one
    double zz = 56;
    zz++; // postincrement (there is also a preincrement operator)
    cout << "zz = " << zz << endl; // => 57
    double yy = -19;
    yy--; // postdecrement (there is also a predecrement operator)
    cout << "yy = " << yy << endl; // => -20

    // floating point numbers
    float r = 1.0;
    float s = 3.0;
    float t = r/s;
    cout << "t = " << t << endl;  // => 0.333333

    // rounding errors
    double xd = 17.0;
    double yd = 3.0;
    double rd = xd/yd;
    cout << "rd = " << rd << endl; // => 5.666667

    double c1 = 1.0;
    double c2 = 2.0;
    double c3 = 3.0;
    double result = c1/c3 + c2/c3;
    cout << "result = " << result; // => 1

    return 0;
}
```

2.5 ASCII Table and Types Conversion

Sometimes we need to force a data type conversion that would not otherwise happen, a process also known as type casting. To type cast a value to a different type, place the desired type in parentheses (...) in front of the value. For data type conversion from integers to `char` and vice versa the ASCII table is used.

The *ASCII table* (American Standard Code for Information Interchange) is a set of 256 codes that many computers use to represent letters, digits, special characters, and other symbols. Only the first 128 of these codes are standardized. The remaining 128 are special characters that are defined by the computer manufacturer.

The *Null character* is denoted by `'\0'`. It is the first element in the ASCII table. The Null character indicates the end of a string (array of characters).

The line

```
int x = (int) 'A';
```

assigns the number 65 to the variable `x`. The line

```
char c = (char) 48;
```

assigns the character `'0'` to the variable `c`. The two lines given above can also be written in short as

```
int x = 'A';
char c = 48;
```

to indicate the type conversion.

As another example, consider

```
int x = 1;
int y = 3;
int r = x/y;  // => 0  (integer division)

double u = (double) x;
double v = (double) y;
double w = u/v;  // => 0.333  (double division)
```

When doing an automatic type conversion, C and C++ converts the lower-ranking (smaller) data item to the higher-ranking (larger) type. The C and C++ language ranks types in this order (with `char` the lowest)

```
char < short < int < long < float < double
```

In addition, `unsigned` types outrank the corresponding `signed` types.

The following program shows an application of the ASCII table.

```
// ASCII.cpp

#include <iostream.h>

int main()
{
   int x = 'A' + 'a';
   cout << "x = " << x << endl;   // => 162

   int r = 'r'/' ';
   cout << "r = " << r << endl;  // => 3 (integer division)

   char c1 = 73;
   char c2 = 32;
   char c3 = 108;
   char c4 = 105;
   char c5 = 107;
   char c6 = 101;
   char c7 = 32;
   char c8 = 67;
   char c9 = 43;
   char c10 = 43;

   cout << c1 << c2 << c3 << c4 << c5 << c6 << c7 << c8 << c9 << c10;
   // => I like C++

   cout << endl;

   int y = 1/4 + ' ' + '0';
   cout << "y = " << y << endl;  // => 80

   int z = '0' + '\0' + 0;
   cout << "z = " << z << endl;  // => 48

   return 0;
}
```

2.6 Precedence Table

Precedence is the priority system for grouping different types of operators with their operands. C, C++ and Java have precedence rules that determine the order of evaluation in expressions that contain more than one operator. When two operators have unequal precedence the operator with higher precedence is evaluated first. Operators with equal precedence are evaluated from left to right. For example, the fourth line of the code fragment

```
int a = 5;
int b = 6;
int c = -1;
int r = a*b + c;    // => 29
int s = a*(b + c); // => 25
```

contains three operations, namely `=`, `*`, `+`. After the precedence table the multiplication operators is applied first, then the addition operator and finally the assignment. Thus result take the value 29. In the expression `a*(b+c)` we first evaluate `b+c` and then multiply with `a`.

The operator `++` is the increment operator, it increments by 1. We have to keep in mind that there is a prefix increment and a postfix increment. They are at different places at the precedence table. A lot of confusion in literature is about the increment and decrement operators, `++` and `--`, respectively. Let `x` be an integer. There is a preincrement operator `++x` and a postincrement operator `x++`. The preincrement operator is at the top of the precedence table and postincrement operator is at the bottom of precedence table. Notice that the postincrement operator is even below the asignment operator `=`.

The operators `*` (multiply), `/` (division) and `%` (remainder) have equal precedence. Operators with equal precedence are evaluated from left to right. All three operators have higher precedence than the assignment operator `=`. For example

```
int m = 27;
int n = 8;
int p = 5;

int r1 = m/n%p;   // => 3
int r2 = n%m/p;   // => 1
```

```
// precedence.cpp

#include <iostream.h>

int main()
{
   int a = 7;
   a++;
   cout << "a = " << a << endl; // => 8

   int b = 5, c = 7;
   int r1 = c*b++;
   cout << "r1 = " << r1 << endl; // => 35

   cout << "b = " << b << endl; // => 6

   int d = 5, e = 7;
   int r2 = d*(e++);
   cout << "r2 = " << r2 << endl; // => 35

   int f = 5, g = 7;
   int r3 = f*++g;
   cout << "r3 = " << r3 << endl; // => 40

   int n = 23, m = 5;

   int s1 = n/m;
   cout << "s1 = " << s1 << endl; // => 4

   int s2 = n%m;
   cout << "s2 = " << s2 << endl; // => 3

   int p = 3;
   int s3 = n/m%p;
   cout << "s3 = " << s3 << endl; // => 1

   return 0;
}
```

2.7 Pointers and References

Pointers are addresses in memory that hold an address. They are used extensively in C and C++ and a thorough understanding of pointers is essential for good programming skills. Typically pointers in C and C++ are typed, i.e. they point to a memory location which contains a certain specified data type. C and C++ support two special address operators: the *address-of-operator* `&` and the *dereference operator* `*` (also called *indirection operator*). The `&` operator returns the address (memory location) of a given variable. If `p` is a pointer to type `int`, then `*p` is the integer to which `p` points.

For example, in the following listing we define an integer `x`. Thus `x` is a memory position (address). The contents at the memory position is the integer 17. Next a pointer `p` to an integer is declared. Then we assign the address of `x` to `p`. The operator `*` also indicates the dereferencing.

```
// pointer1.cpp

#include <iostream.h>

int main()
{
  int x = 17;  // defining an integer variable
               // and assign the value 17

  // to find the memory location of the variable x
  // we do the following
  cout << "address of x is: " << &x << endl;

  int* p = NULL;       // defining a pointer to int
  cout << "address of pointer: " << &p << endl;

  p = &x;              // the pointer p now points to the address of x

  cout << &p << endl;     // address of the pointer p
  cout << p << endl;      // writes the address of x

  cout << *p << endl;     // pushes 17 onto the output stream
                          // the pointer p is dereferenced

  cout << *&p << endl;  // displays address of x (dereferencing once)
  cout << **&p << endl; // displays 17 (dereferencing twice)

  return 0;
}
```

The pointer p is then set to the address of x. We can either print the contents of the pointer variable itself which will be an address or we can dereference the pointer, using the dereferencing operator *. The result of the dereferencing operation is the contents of the memory position to which the pointer p points.

In C and C++ pointers and arrays are closely related as we show more clearly in the next section. For example, C has no predefined string data type. Instead one uses arrays of characters. Newer C++ compilers have a string class.

C++ reference types are closely related to pointer types. *Reference types* create aliases for objects and let us pass arguments to functions by reference. We use the ampersand operator & to indicate a reference. Do not confuse the address-of-operator & which precedes a variable name, with the reference operator &, which follows the data type, for example double& x.

```
// references.cpp

#include <iostream.h>

void main(void)
{
   int i = 17;
   int j = i;
   cout << "i = " << i << endl;   // => 17
   cout << "&i = " << &i << endl; // => address of i
   cout << "j = " << j << endl;   // => 17
   cout << "&j = " << &j << endl; // => address of j
                                  // different from address of i
   int x = 231;
   int& y = x;    // y is an alias of x
   cout << "x = " << x << endl;   // => 231
   cout << "&x = " << &x << endl; // => address of x
   cout << "y = " << y << endl;   // => 231
   cout << "&y = " << &y << endl; // => address of y
                                  // same as address of x
   int k = 17;
   int* kp = NULL;
   kp = &k;
   cout << "*kp = " << *kp << endl; // => 17 (dereference)
   int& m = k;    // reference
   int* mp = NULL;
   mp = &m;
   cout << "kp = " << kp << endl;  // => address
   cout << "mp = " << mp << endl;  // => address same as above
}
```

When we consider cout and pointers to characters we find a different behaviour. This is shown with the following program. Notice that `char* v = NULL;` is also used for string declaration (arrays of characters)

```
// charpoi.cpp

#include <iostream.h>  // for cout
#include <stdio.h>     // for printf

int main()
{

   char c = 'A';
   char* p = NULL;
   p = &c;

   // using cout
   cout << "c = " << c << endl;      // => A
   cout << "*p = " << *p << endl;    // => A
   cout << "p = " << p << endl;      // => A  ???
   cout << "&c = " << &c << endl;    // => A  ???
   cout << "&p = " << &p << endl;    // => 0x00012FF7C
   cout << "(void *)p = " << (void *)p << endl; // => 0x00012FF74

   char d = 'x';
   char* q = NULL;
   q = &d;

   // using printf
   printf("d = %c\n",d);        // => x
   printf("q = %p\n",q);        // => 0012FF70
   printf("&q = %p\n",&q);      // => 0012FF78
   printf("&d = %p\n",&d);      // => 0012FF70
   printf("(void *) q = %p\n",(void *) q); // => 0012FF70

   return 0;
}
```

2.8 Control Statements

2.8.1 Introduction

Control statements control the program flow. For example, selection statements such as `if ... else` and `switch` use certain criteria to select a course of action within a program. Iterative control statements (like `for`, `while`, `do ... while`) on the other hand see to it that under certain conditions control is passed from the last statement in a block to the first statement. The use of these control statements is illustrated in the following sections. To apply the control statements we first have to introduce the *relational operators*. Relational operators allow the user to compare two values yielding a result based on whether the comparison is true or false. If the comparison is false then the resulting value is 0; if true the value is 1. The relational operators in C and C++ are

```
>   greater than
>=  greater than or equal to
<   less than
<=  less than or equal to
==  equal to
!=  not equal to
```

The relational operators can be used for all basic data types. For example

```
unsigned long x = 234567;
unsigned long y = 234568;
if(x == y) ....
```

The relational operators == and `!=` can also be used for pointers. For example

```
double a = 3.14;
double b = 2.78;
double pa = &a;
double pb = &b;
if(pa == pb) ...
```

Selection or flow-control statements select from alternative courses of action by testing certain values. There are two types of selection statements: the `if ... else` and the `switch`.

2.8.2 The if Statement

The `if` statement consists of the `if` keyword followed by a test expression in parentheses, and a single or compound statement. The statement is executed if the test expression is true, or skipped if the expression is false.

The `if` statement has the following pattern

```
if(cond-expression) t-st <else f-st>
```

The *cond-expression* must be of scalar type. The expression is evaluated. If the value is zero (or `NULL` for pointer types), we say that the *cond-expression* is false; otherwise, it is true. If there is no `else` clause and *cond-expression* is true, *t-st* is executed; otherwise, *t-st* is ignored. If the optional `else` *f-st* is present and *cond-expression* is true, *t-st* is executed; otherwise, *t-st* is ignored and *f-st* is executed. C does not have a specific Boolean data type and only newer C++ compiler have a data type `bool`. Any expression of integer or pointer type can serve a Boolean role in conditional tests. The relational expression `(a > b)` (if legal) evaluates to `int` 1 (true) if `(a > b)`, and to `int` 0 (false) if `(a <= b)`. Pointer conversions are such that a pointer can always be correctly compared to a constant expression evaluating to 0. This means, the test for `NULL` pointers can be written

```
if(!ptr) ...
```

or

```
if(ptr == NULL) ....
```

The short cut notation

```
if(ptr) ...
```

stands for

```
if(ptr != NULL) ...
```

The *f-st* and *t-st* statements can themselves be `if` statements, allowing for a series of conditional tests nested to any depth.

```
// ifelse.cpp

#include <iostream.h>

void main(void)
{
   int i;
   cout << "enter an integer: ";
   cin >> i;

   if(i > 5)
   cout << "The number entered is larger than 5" << endl;
   else
   cout << "The number entered is smaller or equal to 5" << endl;

  long m;
  cout << "enter an integer: m = ";
  cin >> m;
  long result = m%2;
  if(result == 0)
  cout << "number is even" << endl;
  else
  cout << "number is odd" << endl;

  int j = -15;
  int* p1 = NULL;
  int* p2 = NULL;

  p1 = &j;
  p2 = &j;

  if(p1 == p2)
  cout << "p1 and p2 are equal" << endl;
  else
  cout << "p1 and p2 are not equal";

  if(&p1 == &p2)
  cout << "&p1 and &p2 are equal" << endl;
  else
  cout << "&p1 and &p2 are not equal" << endl;
}
```

2.8.3 The for Loop, while Loop, do-while Loop

Iteration statements let us loop a set of statements. There are three forms of iteration in C, C++ and Java: `while`, `do - while`, and `for` loops.

The general format for the `while` statement is

```
while(cond-exp) t-st
```

The loop statement, t-st, will be executed repeatedly until the conditional expression, cond-exp, compares to zero (false). The cond-exp is evaluated and tested first. If this value is nonzero (true), t-st, is executed. If no jump statements that exit from the loop is encountered, cond-exp is evaluated again. This cycle repeats until cond-exp is zero.

Pointer type expressions can be compared with the `NULL` pointer. This means that `while(ptr)` is equivalent to

```
while(ptr != NULL) ...
```

The `while` loop offers a concise method for scanning strings (end of string indicate by the null character `'\0'`) and other null-terminated data structures.

The general format for the `do while` statement is

```
do do-st while(cond-exp);
```

The do-st statement is executed repeatedly until cond-exp compares equal to zero (false). The main difference from the `while` statement is that cond-exp is tested after, rather than before, each execution of the loop statement. At least one execution of do-st is assured.

The `for` statement format in C, C++ and Java is

```
for(<init-exp>;<test-exp>;<increment-exp>) statement
```

The sequence of events is as follows. The initializing expression *init-exp*, if any, is executed. This usually initializes one or more loop counters, but the syntax allows an expression of any degree of complexity (including declarations in C and C++). The expression *test-exp* is evaluated following the rules of the `while` loop. If *test-exp* is non-zero (true), the loop statement is executed. An empty expression here is taken as `while(1)`, that is, always true. If the value of *test-exp* is zero (false), the `for` loop terminates. The statement *increment-exp* advances one or more counters. The expression *statement* (possible empty) is evaluated and control returns to the expression *test-exp*.

The while loop is the most general loop and can be used to replace the other two; in other words, a while loop is all we need, and the others are just there for our convenience.

```
// loops.cpp

#include <iostream.h>    // for cout

int main()
{
   int s1 = 0;
   for(int j=1;j<=10;j++)
   {
   s1 = s1 + j;    // can also be written as s1 += j;
   }
   cout << " s1 = " << s1 << "\n";  // => 55

   int i = 0;
   int s2 = 0;
   while(i <= 10)
   {
   s2 = s2 + i;
   i++;
   }
   cout << " s2 = " << s2 << "\n";  // => 55

   int k = 0;
   int s3 = 0;
   do
   {
   s3 += k;
   k++;
   } while(k <= 10);
   cout << " s3 = " << s3 << "\n";  // => 55

   return 0;
}
```

2.8.4 The switch Statement

If one has a large decision tree and all the decisions depend on the value of the same variable we use a `switch` statement instead of a series of `if ... else` constructions. The `switch` statement transfers control to one of several case-labeled statements, depending on the value of the switch expression. Note that if the `break` is omitted, execution will continue over the remaining statements in the switch block. The `break` statement can also be used to break out of an iteration loop.

The `switch` statement uses the following basic format

```
switch(sw-expression) case-st
```

A `switch` statement allows us to transfer control to one of several case-labeled statements, depending on the value of the *sw-expression*. The latter must be of integral type (`char`, `short`, `int`, `long`). In C++ it also can be of class type provided that there is an unambiguous conversion to integral type available. Any statement in *case-st* (including empty statements) can be labeled with one or more case labels.

```
case const-exp-i:case-st-i
```

There can also be at most one `default` label.

```
default:default-st
```

After evaluating `sw-expression`, a match is sought with one of the `const-exp-i`. If a match is found, control passes to the statement `case-st-i` with the matching case label. If no match is found and there is a `default` label, control passes to `default-st`. If no match is found and there is no `default` label, none of the statements in `case-st` is executed. Program execution is not affected when `case` and `default` labels are encountered. Control simply passes through the labels to the following statement or `switch`. To stop execution at the end of a group of statements for a particular case, use `break`.

The program illustrates the syntax of the C++ selection statement `switch`.

```
// switch.cpp

#include <iostream.h>

void main(void)
{
   char st[6] = "bella";

   for(int i=0;i<5; i++)
   {
   char c = st[i];

   switch(c)
   {
   case 'a':  cout << "character is a 'a' " << endl;
     break;
   case 'b':  cout << "character is a 'b' " << endl;
     break;
   default:   cout << "character is not 'a' or 'b' " << endl;
     break;
   }
   }  // end for loop

   int array[] = { 3, 4, 6, 1, 4, -3, 1, 6 };

   for(int j=0; j < 8; j++)
   {
   switch(array[j])
   {
   case 4:  cout << "number at position " << j << " is 4";
     cout << endl;
     break;
   case 6:  cout << "number at position " << j << " is 6";
     cout << endl;
     break;
   default:  cout << "number at position " << j << " is not 4 or 6";
     cout << endl;
     break;
   }
   }  // end for loop

}
```

2.9 Arrays

An *array* is a set of elements with the same data type. The first subscript of a C and C++ array is 0 and the last subscript is 1 less than the number used to declare the array. Thus an array is a contiguous region of storage, large enough to hold all its elements. The first subscript of a C and C++ array is 0 and the last subscript is 1 less than the number used to declare the array.

For example

```
int vector[5];
```

reserves a memory for 5 integer variables which are accessed via

```
vector[0], ... , vector[4] .
```

An array can be declared and initialized in a single statement. For example

```
  int intvec[] = { 1, 78, 12, 5 };
```

allocates memory for four integer variables, initializes these memory positions for the integer values and sets the pointer `intvec` to the start of that memory block. Similarly

```
double matrix[2][2] = { { 1.7, 2.3 },{ 3.2, 7.9 } };
```

declares and initializes a 2 x 2 matrix. The lines

```
double matrix[3][2];
matrix[0][0] = 1.0;  matrix[0][1] = 3.2;
matrix[1][0] = 4.5;  matrix[1][1] = 6.7;
matrix[2][0] = 2.3;  matrix[2][1] = 4.2;
```

declares a 3 x 2 matrix of data type `double` and initializes all the entries.

To access array elements one can use either the index notation or pointer notation. Consider for example

```
int v[5];
*v = 31; *(v+1) = 5; *(v+2) = 9; *(v+3) = 11; *(v+4) = 17;
```

The expression `*(v+j)` has exactly the same effect as `v[j]` and the output is identical. The interpretation of `*(v+j)` is as follows. The name of the array is its address. The expression `v+j` is thus the address with something added to it. Let `j = 3`. Then `v+3` does not mean that we add 3 bytes to `v`. The C++ compiler takes the size of the data into account when it performs arithmetic on data addresses. It knows that `v` is an array of type `int` because it was declared that way. Thus when it sees the expression `v+3` it interprets it as the address of the fourth integer (namely 11) in `v`, not the fourth byte. To find the value at that address we use the dereference operator `*` (also called indirection operator).

```
// array.cpp

#include <iostream.h>

int main(void)
{
   int a[5];
   a[0] = 7; a[1] = 6; a[2] = -67; a[3] = -1; a[4] = 11;
   int product = a[0]*a[1]*a[2]*a[3]*a[4];
   cout << "product = " << product << endl; // => 30954

   double x[4] = { 3.14, 4.7, 5.6, -6.7 };
   double diff = x[2] - x[1];
   cout << "diff = " << diff << endl; // => 0.9

   // using pointer notation to access the element
   double sum = *x + *(x+1) + *(x+2) + *(x+3);
   cout << "sum = " << sum << endl; // => 6.74

   float w[3];
   *w = 3.14; *(w+1) = 1.7; *(w+2) = -0.54;
   cout << "w = " << w << endl;   // starting address
   cout << "*w = " << *w << endl; // => 3.14
   cout << "(w+2) = " << (w+2) << endl; // address of w[2]
   cout << "*(w+2) = " << *(w+2) << endl; // => -0.54

   // copy an array to another
   long u[5];
   u[0] = 4;  u[1] = 14;  u[2] = 16;  u[3] = 19;  u[4] = 21;

   long v[5];

   for(int i=0; i<5; i++)
   {
   v[i] = u[i];
   }
   cout << "v[3] = " << v[3] << endl; // => 19

   return 0;
}
```

Vectors and matrices are, however, best handled by defining an abstract data type (a class) for them and overloading the algebraic and stream operators (see chapter 6).

In the following example program we define a 3 x 3 matrix `M` (two dimensional array). The elements are accessed via `M[i][j]` where `i` and `j` run from 0 to 2. The function `sizeof(M)` returns the size of the entire memory block of 9 double precision floating point numbers. Note that `nrows` and `ncols` are defined as constants. This is essential, because the size of the array must be known at compile time.

```
// statarr.cpp

#include <iostream.h>

void main()
{
  const int nrows = 3;
  const int ncols = 3;

  double M[nrows][ncols];

  cout << "Enter (" << nrows << "x" << ncols << ") matrix M:" << endl;

  for(int nr=0; nr<nrows; nr++)
  {
    for(int nc=0; nc<ncols; nc++)
    {
    cout << "M[" << nr+1 << "," << nc+1 << "] = ";
    cin >> M[nr][nc];
    }
  }
  cout << "sizeof(M) = " << sizeof(M) << endl;
      // returns size of
      //  9 double precision variables, i.e. 9 x 8 = 72
}
```

The above treatment of a two-dimensional array has, however, a severe disadvantage. If one writes routines for matrix multiplication, or for solving the eigenvalue problem, one does not usually know the size of the matrix beforehand. One option would be to define a maximum size and reserve memory for that maximum size irrespective of the actual size required. This would be very wasteful. An alternative approach is to reserve memory at run-time. This is done in C++ via the `new` operator (see section 2.12).

The following program shows an implementation of the farmer, wolf, goat and cabbage problem. The function `safe` returns 1 if the configuration is safe, otherwise zero. For example, the configuration with the farmer and cabbage on the north side and the wolf and goat on the other is unsafe.

```
// Farmer.cpp

#include <iostream.h>

// v[0] describes the state of the farmer
// v[1] describes the state of the wolf
// v[2] describes the state of the cabbage
// v[3] describes the state of the goat
// 0 : south side
// 1 : north side

int safe(int v[])
{
   // goat eats cabbage
   if((v[3] == v[2]) && (v[3] != v[0]))
   return 0;

   // wolf eats goat
   if((v[3] == v[1]) && (v[3] != v[0]))
   return 0;

   // everything else is safe
   return 1;
}

int main(void)
{
   int v[4];
   v[0] = 0; v[1] = 0; v[2] = 0; v[3] = 0;

   int in;

   for(int j=0; j<4; j++)
   {
   cout << "enter the number 0 or 1 : 0 is south, 1 is north";
   cout << endl;
   cout << "v[" << j << "] = ";
   cin >> in;
   v[j] = in;
   }
```

```
   int result = safe(v);
   cout << endl << "result = " << result;

   if(result == 0)
   cout << endl << "configuration is not safe" << endl;
   else
   cout << endl << "configuration is safe" << endl;

   if(v[0] == 0) cout << endl << "farmer is on the south side";
   if(v[0] == 1) cout << endl << "farmer is on the north side";
   if(v[1] == 0) cout << endl << "wolf is on the south side";
   if(v[1] == 1) cout << endl << "wolf is on the north side";
   if(v[2] == 0) cout << endl << "cabbage is on the south side";
   if(v[2] == 1) cout << endl << "cabbage is on the north side";
   if(v[3] == 0) cout << endl << "goat is on the south side";
   if(v[3] == 1) cout << endl << "goat is on the north side";

   return 0;
}
```

2.10 Strings

Strings are arrays of characters. C does not support a separate string data type, but it does provide two slightly different approaches to defining strings. One is to use a character array. The other is to use a character pointer. The line

```
char st0[6];
```

tells the compiler to allocate memory space for up to 5 characters, that is, an array of 5 `char` variables. The 6th space will be filled by the null character `'\0'`. The variable `st0` itself does not contain a character value; it holds the address (some location in memory) cf the zeroth of those 5 `char` variables. To access the characters in a string we use the brackets `[j]`. For the example given above we have `st0[j]`, where `j = 0, 1, 2, 3, 4, 5`. Obviously, `st0[5]` is the null character `'\0'`.

```
// string1.cpp

#include <iostream.h>
#include <string.h>

int main()
{
  char st0[6] = { 'e', 'g', 'o', 'l', 'i', '\0' };
  cout << "st0[0] = " << st0[0] << endl;    // => e
  cout << "st0[4] = " << st0[4] << endl;    // => i
  cout << "&st0 = " << &st0 << endl;        // => address
  cout << "&st0[0] = " << &st0[0] << endl; // => egoli

  char st1[] = "egoli";
  cout << "st1[0] = " << st1[0] << endl;    // => e
  cout << "st1[4] = " << st1[4] << endl;    // => i

  char st2[6];
  strcpy(st2,"Egoli");
  cout << "st2 = " << st2 << endl;     // => Egoli
  cout << "&st2 = " << &st2 << endl;   // => address

  char str3[6] = "egoli";
  cout << "str3 = " << str3 << endl; // => egoli

  return 0;
}
```

The second method one can use to define a string is a character pointer. The asterisk in front of `st0` tells the compiler that `st0` is a pointer to a character. In other words, `st0` holds the address of some character. However, the compiler does not set aside memory space to store characters. However it initializes the pointer to the `NULL` pointer. When the compiler finds the statement

```
st0 = "Egoli";
```

it does two things. First it creates the string `Egoli`, followed by a null character somewhere within the object code file. Then it assigns the starting address of the character `E`.

```
// strings2.cpp

#include <iostream.h>

void main(void)
{
    char* st0 = NULL;
    st0 = "Egoli";

    cout << "st0 = " << st0 << endl;    // => Egoli
    cout << "&st0 = " << &st0 << endl; // => address of pointer

    // to find the address where the string starts
    cout << "(void *) st0 = " << (void *) st0; // => address

    cout << endl;

    cout << "st0[2] = " << st0[2] << endl;  // => o

    // instead of using index notation one can also
    // use pointer notation
    cout << "*(st0+2) = " << *(st0+2) << endl; // => o

    cout << "(st0+2) = " << st0+2 << endl; // => oli
}
```

Thus the line

```
cout << "st0 = " << st0 << endl;
```

does not the address but the string itself. The reason for this is historical – it simulates some functionality of a string data type. If we do the same for an array of integers then things look quite different. We first define a vector of integers which are initialized. We then define a pointer pvec and set it to the starting address of this vector. Now printing out pvec results in the more standard behaviour, i.e. the address of the first element of the vector is printed. Dereferencing the pointer results in the contents of this memory position which is the first element in the array.

Commonly a NULL pointer is used when the address to which the pointer should point is not yet determined (NULL implies undefined).

```
T* p = NULL;
```

Here we defined a pointer to data type T and initialized it to the NULL-pointer.

Similarly, if we want to keep the data type to which the pointer points unspecified we can use a void*. ANSI C allows a pointer of type void* to be assigned to any pointer and also any pointer to be assigned to void*. In C++ we can still assign a pointer of any type to void*, but if we want to assign a void* to any other pointer type we must make an explicit type casting. This is illustrated in the following listing

```
void* pvoid;
char* str;

pvoid = str;              // quite legal in C++
str = pvoid;              // illegal in C++
str = (char*)pvoid        // legal with type-casting
```

We consider strings again when we introduce the new and delete operators.

The string header file string.h includes a large number of useful string manipulation function such as

```
strcmp,  strlen,   strcpy,  strcat
```

The function strcmp compares one string s1 to another s2 character by character (ASCII table). The return data type is int. The function strcmp returns < 0 if s1 is less than s2. If s1 and s2 are the same string it returns 0. If s1 is greater than s2 strcmp returns > 0.

Newer C++ compilers provide a string class. We discuss this string class in chapter 6. The following program shows how the functions strlen, strcmp, strcpy, strcat and strcmpi can be used.

```
// string3.cpp
//
// using functions from string header file

#include <iostream.h>
#include <string.h>

int main()
{
   char str0[6] = "union";
   char str1[17] = "Good Night Egoli";
   char* str2 = "Good-Morning";

   int length1 = strlen(str0);
   cout << "length1 = " << length1 << endl; // => 5

   int length2 = strlen(str1);
   cout << "length2 = " << length2 << endl; // => 16

   int r1 = strcmp(str0,str1); // => 1

   char buffer[100];
   strcpy(buffer,str2);
   strcat(buffer," ");
   strcat(buffer,str1);
   cout << "buffer = " << buffer << endl;

   // strcmpi compares one string with another
   // without case sensitivity
   char* buf1 = "XXX";
   char* buf2 = "xxx";
   int res = strcmpi(buf2,buf1);
   if(res > 0)
   cout << "buf2 is greater than buf1" << endl;
   if(res < 0)
   cout << "buf2 is less than buf1" << endl;
   if(res == 0)
   cout << "buf2 equals buf1" << endl;

   return 0;
}
```

2.11 The sizeof Operator

When we transport a C or C++ program to some other machine, we might want to know how many bytes are in an `int`, since this data type is implementation dependent. The `sizeof` operator yields the number of bytes contained in its operand and can be used with or without parentheses. The `sizeof` operator can also be applied to pointers and arrays. For example, the expression `sizeof(int)` returns the number of bytes that a variable of type integer requires. The values it yields is 4 (OS/2, LINUX, Windows NT) and 2 under DOS.

```
// sizeof.cpp

#include <iostream.h>    // for cout

void main()
{
   int a[4];
   a[0] = 2; a[1] = 7; a[2] = 4; a[3] = -4;
   cout << "&a[0] = " << &a[0] << "\n"; // 0x0003bf80
   cout << &a[1] << "\n";               // 0x0003bf84
   cout << &a[2] << "\n";               // 0x0003bf88
   cout << &a[3] << "\n";               // 0x0003bf8c
   cout << &a[0]+2 << "\n";             // 0x0003bf88

   cout << sizeof(int) << "\n";
   // 4 bytes under OS2, LINUX, Windows NT, 2 bytes under DOS

   cout << sizeof(a) << "\n";
   // 16 bytes under OS2, LINUX, Windows NT, 8 bytes under DOS

   cout << "sizeof(long) = " << sizeof(long) << "\n";  // output 4
   cout << "sizeof(float) = " << sizeof(float) << "\n";// output 4
   cout << "sizeof(double) = " << sizeof(double) << "\n";// output 8
   cout << "sizeof(char) = " << sizeof(char) << "\n";  // output 1
   // 1 byte storage space

   char *str = "willi";
   cout << "sizeof(str) = " << sizeof(str) << "\n";
   // 4 bytes (pointer size)

   int *B = NULL;
   cout << "sizeof(B) = " << sizeof(B) << endl;
   // 4 bytes (pointer size)
}
```

2.12 The new and delete Operators

The `new` operator obtains memory from the operating systems and returns a pointer to its starting point. If the program reserves many chunks of memory using new, eventually the available memory will be reserved and the system will crash. To ensure safe and efficient use of memory, the `new` operator is matched by the corresponding `delete` operator that returns memory to the operating system.

If we want to reserve memory for a vector of `n` integer values we can do this by

```
int *vector = new int[n];
```

This statement declares an integer pointer `vector`, reserves a block of contiguous memory space, large enough for `n` integer variables, and sets the address of the pointer variable `vector` to the start of the memory block.

For a string we use the lines

```
int n = 6;
char* s = NULL;
s = new char[n];
strcpy(s,"Egoli");
```

We must keep in mind that we also have to provide memory space for the *null character* `'\0'` which indicates the end of the string.

Note that `n` need not be constant and can be read in from the terminal. When we no longer require the vector we can free the memory using the `delete` operator and thereby conserve our hardware resources. This is done by the following statement

```
delete[] vector;
```

For allocating memory space for a two-dimensional array we apply the `new` operator twice.

The corresponding C operators to allocate and deallocate memory are `malloc` and `free`. For example

```
char* ptr = NULL;
ptr = (char* ) malloc(6*sizeof(char));
strcp(ptr,"Egoli");
```

The following program shows how one dynamically allocates memory for a one-dimensional array.

```
// newdel.cpp

#include <iostream.h>     // for cout

int main()
{
   int *B = NULL;
   cout << "sizeof(B) = " << sizeof(B) << endl; // => 4
   B = new int[4];

   for(int i=0;i<4;i++)
   { B[i] = i*i + 2;  } // same as *(B+i) = i*i + 2;

   cout << "B[0] = " << B[0] << endl; // => 2
   cout << "*B = " << *B << endl;     // => 2
   // * is the dereference operator

   cout << "B = " << B << endl;    // address
   cout << "&B = " << &B << endl;  // address

   cout << "B[2] = " << B[2] << endl;      // => 6
   cout << "*(B+2) = " << *(B+2) << endl;  // => 6

   cout << "\n" << *(B+2)**(B+3); // => 66
   cout << "\n" << B[2]*B[3];     // => 66
   cout << "\n" << &*(B+1);       // => address
   cout << "\n" << &*(B+2);       // => address

   delete []B;

   int *C = NULL;
   C = new int[6];
   for(i=0;i<6;i++)  {  C[i] = i;  }

   cout << "\n" << C[2]*C[3];  // output 6
   delete []C;

   cout << "\n" << B[1] << "\n";   // 3 ???
   cout << &B[1];                  // address

   return 0;
}
```

The following program shows how one dynamically allocates memory for a two-dimensional array (a matrix).

```
// matrix.cpp

#include <iostream.h>

int main(void)
{
    double** M = NULL;
    int nrows = 3;       // number of rows
    int ncols = 4;       // number of columns
    int nr, nc;

    M = new double* [nrows];
    for(nr=0; nr<nrows; nr++)
    M[nr] = new double[ncols];

    M[0][0] = 3.4; M[0][1] = 1.7; M[0][2] = -5.7; M[0][3] = 7.8;
    M[1][0] = -3.6; M[1][1] = -9.0; M[1][2] = 8.9; M[1][3] = 3.4;
    M[2][0] = 1.1; M[2][1] = 4.5; M[2][2] = 7.8; M[2][3] = 6.7;

    double sum = 0.0;

    for(nr=0; nr<nrows; nr++)
    {
    for(nc=0; nc<ncols; nc++)
    {
    sum += M[nr][nc];
    }
    }
    cout << "sum = " << sum << endl;

    for(nr=0; nr<nrows; nr++)
    delete [] M[nr];
    delete [] M;

    return 0;
}
```

`M` is defined as a pointer to a pointer to a double precision variable. We first reserve memory for `nrows` pointers and set the pointer `M` to the start of that memory block. For each of the `nrows` pointers we reserve enough memory for `ncols` double precision variables.

When using dynamic memory allocation, one should be careful that there are no memory leaks (i.e. that all memory which is allocated during run-time is deallocated again — preferably as soon as it is no longer required) and that the integrity of pointers is ensured (i.e. that pointers do not end up pointing to memory areas which are reserved for other purposes). It is a good programming practice to set all pointers for which there is no memory reserved equal to NULL (known as the NULL pointer).

In the following we show how a three dimensional array can be dynamically allocated. It is left as an exercise to the reader to include the delete operator.

```
// three.cpp

#include <iostream.h>

void main()
{
   int size1 = 2, size2 = 3, size3 = 4;
   int*** A = NULL;
   int i, j, k;

   A = new int **[size1];
   for(i=0; i < size1; i++)
   {
   A[i] = new int *[size2];
   for(j=0; j < size2; j++)
   {
   A[i][j] = new int [size3];
   }
   }

   for(i = 0; i < size1; i++)
   for(j = 0; j < size2; j++)
   for(k = 0; k < size3; k++)
   A[i][j][k] = i + j*j + k*k*k;

   for(i = 0; i < size1; i++)
   for(int j = 0; j < size2; j++)
   for(int k = 0; k < size3; k++)
   cout << "A[" << i << "][" << j << "][" << k << "] = "
        << A[i][j][k] << endl;
}
```

2.13 Arrays of Pointers

An array is a collection of single data units, while a pointer is a special variable that holds the address of a memory location. However, arrays of pointers are not only possible but practical as well. The first part of the following program simply copies five string constants to the two-dimensional array

```
char a[5][15]
```

In making the declaration, the second portion of the subscript determines the maximum string length. Since

```
MICROPROCESSOR
```

is the longest string and it contains 14 characters, the minimum string and that is necessary the array is 15 characters or bytes. This will be just large enough to contain the 14 characters in

```
MICROPROCESSOR
```

plus one more (15 total) for the `NULL` character

```
'\0'
```

The multidimensional array can also be classified as an array of strings or even as an array of char arrays. The array declaration does not allow for dimensioning for different string lengths. If the maximum string length is 15, then storage of this quantity is provided for all five of the string elements. This applies even though most of the strings are less than eight characters in length. Since 15 bytes are provided for each string, there is an obvious waste of memory.

The `strcpy()` function is used to make a copy of the string constants and store each in the bytes provided for each string in the array. Finally, a `for` loop is entered, which causes the contents of the array to be written to the screen. This is a multidimensional array, so the expression `a[i]` is a pointer to the `i` element or `i` string in the array. The array variable name without the subscript brackets was a pointer, while the array name with brackets returned the object. This applies only to single-dimensional arrays. With two dimensions, a single set of brackets with the name is a pointer, while a construct of

```
a[i][j]
```

returns the object to the array position specified by the values of i and j.

Therefore, the expression

```
a[0]
```

where a is a multidimensional array, is a pointer to the first element of the first string in the array. a[1] is a pointer to the first element of the second string, and so on.

In addition to the wasted memory usurped by the array, which must size all string elements to the length of the maximum expected string, there is the fact that storage is required for constants as well as for copies of those constants. This is a most unsatisfactory condition and should be avoided.

Pointers make this a distinct and easily accomplished possibility, as is demonstrated by the second part of program. The second program declares an array of char pointers. The designation

```
char *p[5];
```

declares p to be an array of five char pointers. The pointer to the first string is p[0], the second string p[1], the third p[2], and so on. Nothing is copied to a pointer. Rather, it is assigned a memory address. In this case, each of the five pointers in the array is assigned the address of a constant. This avoids the copying procedure from the previous program and also the doubled storage requirements. The part appears to be identical with the one that use the multidimensional char array.

However, from a memory usage standpoint, the second part is very different. The multidimensional char array version consumed more bytes in object code format and more bytes as an executable file then the pointer version of this same program. This is very significant, and the percentages of savings will grow as program complexity increases, especially in regard to the storage of objects.

An array of strings can also be declared and defined as follows

```
char* name[] = { "Miller", "Smithlonely", "Copper" };
```

Then name[0] corresponds to the string "Miller", name[1] corresponds to the string Smithlonely", and name[2] corresponds to the string Copper.

```
// arrpoi.cpp

#include <iostream.h>   // for cout
#include <string.h>     // for strcpy()

void main()
{
  char a[5][15];
  int i;

  strcpy(a[0], "DATA");
  strcpy(a[1], "LOGIC");
  strcpy(a[2], "MICROPROCESSOR");
  strcpy(a[3], "COMPUTER");
  strcpy(a[4], "DISKETTE");

  for(i=0;i<5;i++)
  {
  cout << "a[" << i << "] = " << a[i] << "\n";
  }
  cout << "a[1][2] = " << a[1][2] << "\n";  // output G
  cout << "a[2][8] = " << a[2][8] << "\n";  // output C
  cout << "a = " << a << "\n\n";  // output 0x8f78ff96 (address)

  char b[5][15] = {
  { "DATA" },
  { "LOGIC" },
  { "MICROPROCESSOR" },
  { "COMPUTER" },
  { "DISKETTE" }
  };

  for(i=0;i<5;i++)
  {
  cout << "b[" << i << "] = " << b[i] << "\n";
  }
  cout << "b[1][2] = " << b[1][2] << "\n";   // output G
  cout << "b[2][8] = " << b[2][8] << "\n\n"; // output C

  char *p[5];
  p[0] = "DATA";
  p[1] = "LOGIC";
  p[2] = "MICROPROCESSOR";
  p[3] = "COMPUTER";
  p[4] = "DISKETTE";
```

```
   for(i=0;i<5;i++)
   {
   cout << "p[" << i << "] = " << p[i] << "\n";
   }
   cout << "p[1][2] = " << p[1][2] << "\n"; // output G
   cout << "p[2][8] = " << p[2][8] << "\n"; // output C
   cout << "p = " << p << "\n";  // output 0x8f7fffec (address)

   char *q[5] = {
   { "DATA" },
   { "LOGIC" },
   { "MICROPROCESSOR" },
   { "COMPUTER" },
   { "DISKETTE" }
   };

   for(i=0;i<5;i++)
   {
   cout << "q[" << i << "] = " << q[i] << "\n";
   }
   cout << "q[1][2] = " << q[1][2] << "\n";    // output G
   cout << "q[2][8] = " << q[2][8] << "\n\n";  // output C

   char* name[] = { "Miller", "Smithlonely", "Copper" };

   cout << name << endl;    // address
   cout << &name << endl;   // same address
   cout << name[0] << endl; // => Miller
   cout << name[1] << endl; // => Smithlonely

   cout << (void*) *name << endl;
   cout << (void*) &*name[0] << endl;
   cout << (void*) &*name[1] << endl;
   cout << (void*) &*name[2] << endl;

   cout << name[1][3] << endl; // => t
   cout << name[2][1] << endl; // => o
}
```

2.14 Logical AND, Logical OR and Logical NOT

The logical operators in C, C++ and Java are:

```
&&  logical AND
||  logical OR
!   logical NOT
```

The logical operators work with logical values (true and false) allowing us to combine relational expressions. The logical operators always produce a result of either 0 (false) or 1 (true). The logical operators `&&` and `||` are short circuit. Suppose that we have the expression

```
exp1 && exp2
```

If `exp1` is false, then the entire expression is false, so `exp2` will never be evaluated. Likewise given the expression

```
exp1 || exp2
```

`exp2` will never be evaluated if `exp1` is true. As an example consider

```
int x = -5;
int y = 7;
int result = ((x > 0) && (y > 0));
```

The first expression is false thus the second expression is not evaluated.

Warning. Look out for the operator precedence. For example the relational operators (`<` `>` `==`) have higher precedence than the logical AND operator. Note that

```
if(!n)
```

is equivalent to

```
if(n==0)
```

Note that the bitwise AND operator is `&` and the bitwise OR operator is `|`

The logical operators should not be confused with the bitwise operators `&`, `|`, `~`.

```
// logical.cpp

#include <iostream.h>

int main()
{
   cout << endl;
   int j;
   cout << "enter an integer: ";
   cin >> j;

   if(((j%2) == 0) && (j < 10))
   cout << "the integer is even and smaller than 10";
   else
   cout << "the integer is either odd or larger than 10 or both";

   cout << endl;
   int k;
   cout << "enter an integer: ";
   cin >> k;

   if((k > 0) || (k < 0))
   cout << "the integer is nonzero";
   else
   cout << "the integer is zero";

   cout << endl;
   int n;
   cout << "enter an integer: ";
   cin >> n;

   if(!n)          // equivalent to  n == 0
   cout << "the integer is zero";
   else
   cout << "the integer is nonzero";

   return 0;
}
```

2.15 Pass by Value, Pass by Reference

Arguments to functions can be passed either by value or by reference. When an argument is passed by value, a copy of the argument is produced, and the associated parameter is the same as a local variable for the function. This means, changes to the parameter variable will have no effect on the original argument value. Functions that receive variables as parameters get local copies of those variables, not the originals. The alternative, pass by reference, is indicated by the presence of the ampersand operator & in the argument list. When arguments are passed by reference, the parameter variable is an alias for the argument value. Thus, changes in the parameter also alter the original argument. In C and obviously in C++ we also can use pointers and the dereference operator to pass by reference. Thus functions that receive pointers to variables gain access to the original variables associated with the pointers.

In the following program we pass the first argument of the function f by value and the second by reference.

```
// pass1.cpp

#include <iostream.h>

void f(double x, double& y, double* z)
{
  y = x*x;      // y <- 2*2 = 4
  x++;          // x <- x+1 = 3
  *z = y + 3; // z <- 7
}

void main()
{
  double x = 2, y = 0;
  double z;
  f(x,y,&z);

  cout << "y = " << y << endl;     // => 4
  cout << "x = " << x << endl;     // => 2 !!! (pass by value)
  cout << "z = " << z << endl;     // => 7
}
```

It is often inefficient to pass large structures (e.g. large matrices or large records) by value. Instead one can pass them by reference and declare them constant const.

```
void f(const matrix& A)
```

A common pitfall for beginners is the swapping of two basic data types, for example integers. In the following program we show how two integer numbers can be swapped using passing by reference. We also included the pitfall (pass by value). We also included the addresses to show that in passing by value the addresses in `main` and in the function `sw1` are different.

```
// pass2.cpp

#include <iostream.h>              // for cout

void sw1(int a, int b)             // pitfall
{
   int t;
   cout << "&a = " << &a << endl;  // 0x0003bf70
   t = a; a = b; b = t;
   cout << "&a = " << &a << endl;  // 0x0003bf70
}

void sw2(int *px, int *py)         // C version (pointer)
{
   int t;
   cout << "px = " << px << endl;  // 0x0003bf88
   t = *px; *px = *py; *py = t;
   cout << "px = " << px << endl;  // 0x0003bf88
}

void sw3(int &u, int &v)           // C++ version (reference parameter)
{
   int t;
   cout << "&u = " << &u << endl;  // 0x0003bf80
   t = u; u = v; v = t;
   cout << "&u = " << &u << endl;  // 0x0003bf80
}

int main()
{
   cout << endl << endl;
   int a1, b1;
   a1 = 7; b1 = 3;
   cout << "&a1 = " << &a1 << endl;            // 0x0003bf8c

   sw1(a1,b1);
   cout << "a1 = " << a1 << endl;              // 7
   cout << "b1 = " << b1 << endl;              // 3
```

```
  cout << "&a1 = " << &a1 << endl << endl;  // 0x0003bf8c

  int x1, y1;
  x1 = 13, y1 = 9;
  cout << "&x1 = " << &x1 << endl;          // 0x0003bf88

  sw2(&x1,&y1);
  cout << "x1 = " << x1 << endl;            // 9
  cout << "y1 = " << y1 << endl;            // 13
  cout << "&x1 = " << &x1 << endl << endl;  // 0x0003bf88

  int u1, v1;
  u1 = 1; v1 = -1;
  cout << "&u1 = " << &u1 << endl;          // 0x0003bf80

  sw3(u1,v1);
  cout << "u1 = " << u1 << endl;            // -1
  cout << "v1 = " << v1 << endl;            // 1
  cout << "&u1 = " << &u1 << endl << endl;  // 0x0003bf80

  return 0;
}
```

The temporary variable `t` inside the functions `sw2` and `sw3` can be avoided by using the bitwise XOR operation. We have

```
int a = 5;
int b = 13;
a = a^b;
b = b^a;
a = a^b;
cout << "a = " << a << endl; // => 13
cout << "b = " << b << endl; // => 5
```

2.16 Command-Line Arguments

C++ enables us to access command-line arguments by supplying and using the following parameters in function `main`:

```
main(int argc, char* argv[])
```

`argc` is the traditional name for the first argument to the main function in a C source program. It is an integer that specifies how many arguments are passed to the program from the command line. `argv` is the traditional name for the second argument to the main function in a source program. It is a pointer to an array of strings. Traditionally, the first string is the program name, and each following string is an argument passed to the program from the command line.

The `argc` parameter returns the number of command-line arguments. The `argv` parameter is a character pointer that accesses the various command-line arguments. The value of argc takes into account the name of the program itself. The expression `argv[0]` is the pointer to the program's name. The expression `argv[1]` is a pointer to the first argument, and so on. Compile to the execute file

```
calculus.exe
```

At the command prompt type

```
calculus number1 operator number2
```

For example

```
calculus 3.0 / 4.0
```

There must be a whitespace between the number1 and the operator as well as between the operator and number2.

The following program implements a simple five function calculator

addition, subtraction, multiplication, division, exponentation

The exponentiation is given by `p`, for example

```
calculus 2 p 4
```

evaluates the expression as 16.

```
// calculus.cpp

#include <iostream.h>  // for cout
#include <stdlib.h>    // for exit(), for atof()
#include <string.h>    // for strcpy()
#include <math.h>      // for pow()

int main(int argc, char* argv[])
{
   char op;
   int error;
   char st[2];
   double r, f, s;

   if(argc < 4) {
      cout << "Proper arguments : <number><operation><number>";
      cout << "\n\n";
      exit(0);
 }

 f = atof(argv[1]);  s = atof(argv[3]);
 strcpy(st,argv[2]);  op = st[0];

 if(op != '+' && op != '-' && op != '*' && op != '/' && op != 'p')
   error = 1;
   else
   error = 0;

   if(error == 1) { cout << "bad operator";  exit(0); }

   if(op == '/' && s == 0.0)
   { cout << "division by 0 not allowed";  exit(0); }

  switch(op) {
    case '+' :   r = f+s;  break;
    case '-' :   r = f-s;  break;
    case '*' :   r = f*s;  break;
    case '/' :   r = f/s;  break;
    case 'p' :   r = pow(f,s);  break;
    }
    cout << "result = " << r;

    return 0;
  }
```

In the following program we have a function `topolar(...)` which receives the rectangular coordinates by value (since they are not altered) and the polar coordinates `r`, `theta` and `phi` by reference. The function `main` has two arguments, an integer variable `argc` which holds the number of command line parameters and an array of strings `argv[]` which holds the actual command line parameters. If we start the program from the command line by typing

```
topolar 1.0 3.0 1.2 <ENTER>
```

`argc` will be set to three and

```
argv[1] = "1.0",  argv[2] = "3.0", argv[3] = "1.2" .
```

The `argv[0]` will hold the name of the program (`"topolar.exe"`). The `switch` statement checks for the number of command-line parameters. If there are 3 command-line parameters it uses these for the rectangular coordinates. If there are no command-line parameters, the rectangular coordinates are read from the standard input stream (the keyboard), otherwise the program aborts with an error message.

```
// topolar.cpp

#include <iostream.h>    // for cout and cin
#include <stdlib.h>      // for exit()
#include <math.h>        // for sin, acos, sqrt, atof

void toPolar(double x,double y,double z,
             double& r,double& theta,double& phi)
{
       // x, y and z do not change and are sent as variables
       // (a local copy is made)
       // r, theta and phi are to be changed and are hence sent
       // via reference (the same memory area is used as in the
       // calling program)

  r = sqrt(x*x+y*y+z*z);

  if(r != 0.0)
    theta = acos(z/r);
  else
    theta = 0.0;

  if((theta != 0.0) && (r != 0.0))
    phi = acos(x/(r*sin(theta)));
  else
    phi = 0.0;
}
```

```
void readCoordinates(double& x,double& y,double& z)
{
  cout << "Enter x coordinate : ";  cin >> x;
  cout << "Enter y coordinate : ";  cin >> y;
  cout << "Enter z coordinate : ";  cin >> z;
}

void main(int argc, char* argv[])
{
// argc holds the no of command line parameters and
// the array of strings argv[] holds the command line
// parameters itself. The name of the program is
// passed in argv[0]

  double x, y, z;

  cout << "::" << argv[0] << "::" << endl;

  switch(argc-1)
  {
    case 0: readCoordinates(x,y,z);  // when no command line
    break;                           // parameters, read x,y,z

    case 3:  x = atof(argv[1]);  // converts first string to double
     y = atof(argv[2]);
     z = atof(argv[3]);
     break;

    default: cout << "*** ERROR *** : invalid number of command "
  << "line parameters." << endl;
     exit(0);
  }

  double r, theta, phi;

  toPolar(x,y,z,r,theta,phi);  // calculate polar coordinates

  cout << "(" << x << "," << y << "," << z << ")  => " << r
       << " r  + " << theta << "   + " << phi << " " << endl;
}
```

2.17 Bitwise Operations

The *bitwise operation* in C, C++ and Java are: bitwise AND `&`, bitwise OR `|`, bitwise XOR `^` and bitwise NOT `~`. To understand the bitwise operation we have to use the fact that every integer number can be written in binary notation. For example the integer number 14 (i. e. base 10) can be written as

$$14 = 1 \cdot 2^3 + 1 \cdot 2^2 + 1 \cdot 2^1 + 0 \cdot 2^0 \,.$$

Thus in binary we can express 14 (base 10) as bitstring 1110. The bitwise operations are as follows.

```
bitwise AND &
1 1 => 1
0 1 => 0
1 0 => 0
0 0 => 0

bitwise inclusive OR |
1 1 => 1
1 0 => 1
0 1 => 1
0 0 => 0

bitwise exclusive OR ^ (XOR)
1 1 => 0
1 0 => 1
0 1 => 1
0 0 => 0

bitwise NOT operator ~ (one's complement)
0 => 1
1 => 0
```

As an example consider the two (decimal) numbers 14 and 9. In binary 9 is given by 1001. Thus the bitwise AND of these integers is

```
00000000 00000000 00000000 00001001
00000000 00000000 00000000 00001110
-----------------------------------
00000000 00000000 00000000 00001000
```

here we assumed that the register on the CPU which holds the bit string is 32 bits long. The result of the bitwise AND operation is 8 in decimal.

```
// bitwise.cpp

#include <iostream.h>

int main()
{
   int r3 = 4; // binary 100b
   int r4 = 5; // binary 101b
   int r5 = r3 & r4;   // bitwise AND
   cout << r5 << endl; // => 4 => binary 100b

   int r6 = 7;  // binary  111b
   int r7 = 11; // binary 1011b
   int r8 = r6 | r7;   // bitwise OR
   cout << r8 << endl; // => 15 => binary 1111b

   int r9 = r6 ^ r7;   // bitwise XOR
   cout << r9 << endl; // => 12 => binary 1100b

   int x = 125;
   int r10 = x ^ x;
   cout << "r10 = " << r10 << endl; // => 0

   int r11 = ~r9;
   cout << r11 << endl; // => -13

   // two complement
   // to find negative number of a positive number
   // first we do one's complement and then we add 1
   // using the preincrement operator
   int r12 = 4;
   int r13 = ~r12;   // one's complement
   int r14 = ++r13;
   cout << r14 << endl; // => -4

   return 0;
}
```

2.18 Shift Operations

The *shift operators* are `<<` and `>>`. The left shift operator is `<<` and the right shift operator is `>>`. In the expression `E1 << E2` and `E1 >> E2`, the operands `E1` and `E2` must be of integral type. The normal integral type conversions are performed on `E1` and `E2`, and the type of the result is the type of the type converted `E1`. If `E2` is negative or is greater than or equal to the width in bits of `E1`, the operation is undefined. The result of `E1 << E2` is the value `E1` left-shifted by `E2` bit positions, zero-filled from the right if necessary. Left shift of an `unsigned long` `E1` are equivalent to multiplying `E1` by 2^{E1}, reduced modulo `ULONG_MAX+1`; left shifts of `unsigned ints` are equivalent to multiplying by 2^{E2} reduced modulo `UINT_MAX`. The result of `E1 >> E2` is the value of `E1` right-shifted by `E2` bit position. If `E1` is of unsigned type, zero-fill occurs from the left if necessary. If `E1` is of signed type, the fill from the left uses the sign bit (0 for positive, 1 for negative `E1`). This sign-bit extension ensures that the sign of `E1 >> E2` is the same as the sign of `E1`. Except for signed types, the value of `E1 >> E2` is the integral part (integer division) of the quotient $E1/2^{E2}$.

```
// myshift.cpp

#include <iostream.h>

int main()
{
   int r1 = 2 << 4;  // 2 = 10b (binary)
   // shift left 4 => 100000b => 32
   cout << r1 << endl; // => 32

   int r2 = 2 << 31;
   // shift left 31 => 1 00000000000000000C00000000000000b
   cout << r2 << endl; // => 0

   int r3 = 15 >> 2; // 15 = 1111b
   // shift right 2 => 11b => 3
   cout << r3 << endl; // => 3

   return 0;
}
```

```
// mybitset.cpp

#include <iostream.h>

inline void setbit(unsigned long& x,unsigned long b)
{
   x |= (1 << b);
}

inline void clearbit(unsigned long& x,unsigned long b)
{
   x &= ~(1 << b);
}

inline unsigned long testbit(unsigned long x, unsigned long b)
{
   return ((x & (1 << b)) != 0);
}

int main(void)
{
   unsigned long b = 3;
   unsigned long x = 7;  // binary 111
   setbit(x,b);
   cout << "x = " << x << endl; // => 15 decimal => 1111 binary
   clearbit(x,b);
   cout << "x = " << x << endl; // => 7 => binary 111
   unsigned long r = testbit(x,b);
   cout << "r = " << r << endl; // => 0

   unsigned long y = 17;
   setbit(y,b);
   cout << "y = " << y << endl; // => 25
   clearbit(y,b);
   cout << "y = " << y << endl; // => 17
   unsigned long s = testbit(y,b);
   cout << "s = " << s << endl; // => 0

   unsigned long z = 8; // 8 => 1000 binary
   unsigned long t = testbit(z,b);
   cout << "t = " << t << endl; // => 1

   return 0;
}
```

2.19 Passing a Function to a Function

Like other pointers, function pointers can be passed as arguments. For example, the statement

```
double (*func)(double);
```

declares a pointer to a function returning a double and accepting a single double parameter. The program passftpf.cpp demonstrates how to pass a function as an argument.

```
// passftf.cpp
//
// recall that sin(0) = 0;  cos(0) = 1;  exp(0) = 1;

#include <iostream.h>
#include <math.h>      // why do we need this header file ?

double func1(double x) {  return (exp(x) + x + 1); }

double func2(double x) {  return (sin(x) + cos(x) + 2 + x); }

double func3(double x) { return (pow(x,3) + pow(x,2) + x + 1); }

// f is a pointer to the function
double map(double (*f)(double),double* v) {  return (f(*v)); }

void main(void)
{
   double res1;
   double y = 0.0;
   res1 = map(func1,&y);
   cout << "res1 = " << res1 << endl;  // => 2

   double res2;
   double w = 0.0;
   res2 = map(func2,&w);
   cout << "res2 = " << res2 << endl;  // => 3

   double res3;
   double z = 2.0;
   res3 = map(func3,&z);
   cout << "res3 = " << res3 << endl;  // => 15
}
```

2.20 Recursion

Recursion plays a central role in computer science. A recursive function is one whose definition includes a call to itself. A recursion needs a stopping condition. Of course, it is not allowed to use the `main` function in a recursive call.

We consider three examples. In the first example we show how multiplication can be implemented using recursion. The second example shows an implementation of division using subtraction and recursion. The length of a string can be found using recursion. At every step we have to test whether we reached the end of the string indicated by the null character `'\0'`.

```
// recursion1.cpp
// multiplication of two numbers using recursion

#include <iostream.h>

unsigned long mult(unsigned long a,unsigned long b)
{
   unsigned long result = 0;
   if(a == 0) return 0;
   if(b == 0) return 0;
   if(a == 1) return b;
   if(b == 1) return a;
   else
     result = mult(a,b-1) + a;
   return result;
}

int main()
{
   unsigned long n, m;
   cout << "enter a non-negative integer n = ";
   cin >> n;
   cout << "enter a non-negative integer m = ";
   cin >> m;

   unsigned long result = mult(n,m);
   cout << "result = " << result << endl;

   return 0;
}
```

```
// recursion2.cpp
//
// integer division of two numbers using recursion
// b is the divisor
// a is the dividend

#include <iostream.h>
#include <stdlib.h>

unsigned long divide(unsigned long a,unsigned long b)
{
   unsigned long result = 0;
   if(b > a) return 0;
   if(b == 0) { cout << "division by 0 not allowed"; exit(0); }
   if(b == 1) return a;
   else
   result = divide(a-b,b) + 1;

   return result;
}

int main()
{
   unsigned long n, m;
   cout << "enter a non-negative integer n = ";
   cin >> n;
   cout << "enter a non-negative integer m = ";
   cin >> m;

   unsigned long result = divide(n,m);
   cout << "result = " << result << endl;

   return 0;
}
```

The program determines the length of a string recursively A string consists of contiguous characters in memory, ending with the `NULL` character `'\0'`. Conceptually, we can think of a string as either the `NULL` string, consisting of just the `NULL` character, or as a character followed by a string. This definition of a string describes it as a recursive data structure. Thus we can use this to code basic string handling functions recursively.

```
// recursion3.cpp

#include <iostream.h>
#include <string.h>    // for strcpy

int length(char *s)
{
   if(*s == '\0')
      return 0;
   else
      return (1 + length(s+1));
}

int main()
{
   char* st = "willi hans";
   int res1 = length(st);
   cout << "The length of the string st is: " << res1; // => 10
   cout << endl;

   char* empty = "\0";
   int res2 = length(empty);
   cout << "The length of the string empty is: " << res2; // => 0
   cout << endl;

   char* z = NULL;
   z = new char[4];      // allocating memory
   strcpy(z,"oli");
   int res3 = length(z);
   cout << "The length of the string z is: " << res3; // => 3
   delete [] z;

   return 0;
}
```

2.21 Inline Assembly Language

Assembly language is nothing more than a human form of the language of the computer itself. Assembly language is very powerful, it is the only way to tap the full power of the CPU. In our example the CPU is the Pentium (586). Most C++ compilers allow inline assembler language. `EAX`, `EBX` and `ECX` are 32 bit registers on the CPU (Pentium). The 32 bit `EBP` register is the base pointer. It can be used as a memory pointer. `EBP` points relative to the stack segment register. It is the square bracket `[ ]` that indicates that the memory location pointed to by `EBP`, rather than `EBP` itself, is the source operand. Thus `[EBP]` is the contents in memory (stack).

```
// assemb.cpp

#include <iostream.h>

void main()
{
   long a = 67;
   long b = 123;
   long c = a + b;
   cout << "c = " << c <<endl;

   long r;
   // inline assembly language (Microsoft style)
   _asm
   {
   MOV EAX, 67
   MOV EBX, 123
   ADD EBX, EAX
   MOV r, EBX
   }
   cout << "r = " << r << endl;

   long s;
   _asm
   {
   MOV dword ptr [EBP-8], 67
   MOV ECX, 123
   ADD ECX, dword ptr [EBP-8]
   MOV s, ECX
   }
   cout << "s = " << s << endl;
}
```

2.22 Time Header File

The time functions allow us to obtain the current time, then convert and store it. The current time is always taken from the system time. The function

```
char* ctime(const time_t* timer)
```

returns a pointer to the character string result; `NULL` if `timer` represents a data before 1980. It converts it into a 26-character string in the following form, terminating with a newline character and a null character

```
Sat Apr 01 14:09:43 1999\n\0
```

All the fields have constant width. The function `difftime` finds the difference between two times.

```
// mytime.cpp

#include <iostream>
#include <time.h>      // for ctime, difftime
#include <string.h>
#include <stdlib.h>
#include <string>
using namespace std;

int main()
{
   time_t t;
   t = time(NULL);
   cout << ctime(&t) << endl;

   char* st = NULL;
   st = new char[26];
   strcpy(st,ctime(&t));
   cout << st;

   string s = string(ctime(&t));
   cout << "s = " << s << endl;

   char* weekday = new char[4];
   int i;
   for(i=0; i<3; i++)
   weekday[i] = st[i];
   cout << "weekday = " << weekday << endl;
```

```
char* month = new char[4];
for(i=4; i<7; i++)
month[i-4] = st[i];
cout << "month = " << month << endl;

char* day = new char[3];
for(i=8; i<10; i++)
day[i-8] = st[i];
cout << "day = " << day << endl;

char* year = new char[5];
for(i=20; i<24; i++)
year[i-20] = st[i];
cout << "year = " << year << endl;

int nyear = atoi(year);
cout << "nyear = " << nyear << endl;

// using difftime
unsigned long k, l;
unsigned long m, n;
time_t first, second;

first = time(NULL);

for(k=0; k<20000; ++k)
{
m = 0; n = 0;
for(l=0; l<20000; ++l)
{
m++; n++;
}
}
second = time(NULL);

double difference = difftime(second,first);
cout << "difference in time: " << difference;

return 0;
}
```

2.23 Random Numbers

ANSI C and ANSI C++ includes an integer random number generator. The function `srand` initializes the random number generator. The syntax is

```
void srand(unsigned seed);
```

The function `rand` uses a multiplicative congruential random number generator with period 2^{32} to return successive pseudorandom numbers in the range from 0 to `RAND_MAX`. Then `rand` returns the generated pseudo-random number. `rand` is available on UNIX systems and is compatible with ANSI C and ANSI C++.

```
// random.cpp

#include <iostream.h>
#include <stdlib.h>  // for srand, rand
#include <time.h>    // for time_t

int main()
{
   time_t t;
   srand((unsigned) time(&t)); // seed

   cout << "Ten random numbers from 0 to 99" << endl;
   int i;
   for(i=0; i<10; i++)
   cout << rand()%100 << endl;

   double x = (double) rand();
   double y = (double) RAND_MAX;
   cout << x/y << endl;

   cout << "Random binary string " << endl;
   char* bstr = NULL;
   bstr = new char [33];
   for(i=0; i<32; i++)
   bstr[i] = rand()%2;

   return 0;
}
```

2.24 Constants and Inline Functions

For C-programmers it was common to define constants using macros. For example, one would define

```
#define eps 1.0e-8
```

In this case the C or C++ preprocessor replaces every occurrence of `eps` with `1.0e-8` before the source code is forwarded to the compiler. This has several disadvantages. Firstly any compiler error involving `eps` will refer to `1.0e-8` and not to `eps`. If the code is long it might take the programmer a very long time before he finds the error. The same problem is encountered when one uses a symbolic debugger, since `eps` is never entered into the programs symbol table. Furthermore, the constant `eps` cannot be given an explicit type.

All these problems can be solved in C++ by replacing the above macro by

```
const double a = 1.0e-8;
```

The constant can be defined either locally or globally (preferably locally). An even bigger sin is to use macros in order to define *inline functions*. Consider the following commonly used macros

```
#define SQR(x) x*x

#define MAX(x,y) ((x) > (y) ? (x) : (y))
```

One needs the parenthesis around `x` and `y` in order to allow the user to use the macro with expressions. The macros look harmless enough. Consider, however, the following pitfalls

```
y = SQR(1+2);      // result: 5 instead of 9

z = MAX(x++,y);
```

In the first case the macro expansion results in

```
y = 1+2 * 1+2;
```

which evaluates to 5 (since multiplication has higher precedence than addition). In the second case `x` will be incremented either once or twice, depending on whether `(x+1)` is greater than `y` or not. Macros like those above are not only bad style but also downright dangerous. Again there is an elegant alternative available in C++, namely using function templates (see chapter 5)

```
// nomacro.cpp

#include <iostream.h>

  template <class T>
  inline T sqr(T x)
  {
  return x*x;
  }

  template <class T>
  inline T& max(T& x, T& y)
  {
    if(x > y)
    return x;
    else
    return y;
  }

int main()
{
   cout << "sqr(k) = " << sqr(2+1) << endl; // => 9

   double a = 2.1;
   double b = 1.7;

   cout << "max(a,b) = " << max(a,b) << endl; // => 2.1

   return 0;
}
```

Declaring these functions as `inline` avoids the function call overheads. Note, however, that `inline` is a request to the compiler which might be ignored if the compiler feels it is not a good idea (e.g. for example for recursive functions and complex functions).

2.25 Jump Statements

A jump statement, when executed, transfers control unconditionally. There are for such statements:

```
break,  continue,  goto,  return
```

A `break` statement can be used only inside an iteration (`while`, `do-while` and `for` loops) or `switch` statement. It terminates the iteration or `switch` statement. Since iteration and switch statements can be intermixed and nested to any depth, we have to take care that the `break` exits from the correct loop or switch. The rule is that a `break` terminates the nearest enclosing iteration or `switch` statement.

The syntax for the `continue` statement is

```
continue
```

A `continue` statement can be used only inside an iteration statement. It transfers control to the test condition for `while` and `do while` loops, and to the increment expression in a `for` loop. With nested iteration loops, a `continue` statement is taken as belonging to the nearest enclosing iteration.

The syntax for the `goto` statement is

```
goto Label
```

The `goto` statement transfers control to the statement labeled Label, which must be in the same function. One good use for the `goto` is to exit from a deeply nested routine. A simple `break` statement would not work here, because it would only cause the program to exit from the innermost loop.

Unless the function return type is `void`, a function body must contain at least one return statement with the following format

```
return return-expression;
```

where return-expression must be of type `type` or of a type that is convertible to `type` by assignment. The value of the return-expression is the value returned by the function.

```
// didact.cpp

#include <iostream.h>
#include <stdlib.h>
#include <time.h>

int main()
{
  time_t t;
  srand((unsigned long) time(&t)); // seed

  L3:
  int a = rand()%100;
  int b = rand()%100;
  int result;
  L1:
  cout << a << " + " << b << " = ";
  cin >> result;
  if(result == (a+b))
  goto L2;
  cout << "sorry you are not correct: try again" << endl;
  goto L1;
  L2:
  cout << "congratulations you are correct" << endl;
  char c;
  cout << "Do you want to add again: Press y for yes and n for no: ";
  cin >> c;

  if(c == 'y') goto L3;
  if(c == 'n')
  {
  cout << "bye, see you next time around";
  goto L4;
  }
  L4:

  return 0;
}
```

2.26 The static Keyword

Static declarations have two important uses. The first and more elementary use is to allow a local variable to retain its previous value when the block or function is reentered. This is in contrast to an automatic variable, which loses its value upon exit and must be reinitialized upon reentry. An example is shown in the following code. The second and more subtle use of `static` in external declarations is that it provides a privacy mechanism that is very important to maintaining the modularity of programs. The term privacy refers to the visibility or scope restrictions on the accessibility of variables and functions. Static external declarations are visible only within the file in which they are defined. Thus, unlike external variables that can be accessed by other files, a static declaration is available only throughout its own file. In C++, external variables and static variables are initialized to zero if they are not explicitly initialized by the programmer. In contrast, automatic and register variables are not initialized by the system and can start with garbage values. For object-oriented programming, static should be used only as a privacy mechanism.

```
// mstatic.cpp

#include <iostream.h>

int min(const int x,const int y)
{
   static unsigned long counter = 0;
   counter++;
   cout << "min is called " << counter << " times";
   cout << endl;
   if(y < x) return y;
   else
   return x;
}

void main(void)
{
   int a = 3, b = 7;
   int result = min(a,b);
   cout << "result = " << result << endl;

   int i = 15, j = -12;
   result = min(i,j);
   cout << "result = " << result << endl;
}
```

2.27 Structures

Like an array, a structure is a collection of related data items under one name. While array elements are all of the same data type, the elements of a structure, known as members, can be of different data types. This ability to group different data types in the same construct makes structures very powerful and flexible data-handling tools. For example

```
struct Person
{
   char name[12];
   char birthdate[9];
   int idnumber;
   char phone[10];
}
```

Before we can use a structure variable, we must declare a structure type. A structure-type declaration begins with the keyword `struct`. Declaration of the structure's member appear in curly braces after the keyword. A declaration like the one above creates a template that structure variables of the defined type will follow. Thus we created a new data type (sometimes call abstract data type) called `Person`. In this case `Person` is the structure tag. Once we have declared a structure, we can create variable of that data type. For example

```
struct Person p;
```

or

```
struct Person p[4];
```

if we need an array of the data type `Person`. The members of a structure are identified by name, using the *member-of-operator* . . This operator is sometimes also called access operator. For example, we have

```
strcpy(p.name,"Smith, A");
p.idnumber = 999;
```

```
// structure.cpp

#include <iostream.h>
#include <string.h>

struct Invoice
{
  char client[30];
  int acctnum;
  double amount;
};

int main()
{

  struct Invoice x[3];

  strcpy(x[0].client,"Miller, A.");
  x[0].acctnum = 17;
  x[0].amount = 7777.11;

  strcpy(x[1].client,"Smith, R.");
  x[1].acctnum = 18;
  x[1].amount = 234.11;

  strcpy(x[2].client,"Coppersmithlone, L. R.");
  x[2].acctnum = 21;
  x[2].amount = 87.11;

  for(int i=0; i<3; i++)
  {
  cout << "client: " << i << " " << x[i].client << " "
       << x[i].acctnum << " " << x[i].amount << endl;
  }

  int size = sizeof(x);
  cout << "size = " << size << endl; // =>

  return 0;
}
```

Next we consider structures and pointers. The declaration

```
struct Employee q, *p;
```

names `q` a variable of type `Employee` and `p` is a pointer of the same type. The line

```
p = &q;
```

assigns `p` the address of `q`. Pointer `p` now points to the storage address reserved for `struct Employee`. The structure pointer operator `->` is used to access each structure element. This special operator is simply a convenient shorthand method to access when dealing with pointers and structures (or classes). The expression

```
p -> name
```

means exactly the same as

```
(*p).name
```

The latter uses the structure member operator `.` in conjunction with the unary `*` operator for access. The parentheses are necessary because the struture member operator `.` takes higher precedence than does the unary `*` operator.

```
// employee.cpp

#include <iostream.h>
#include <string.h>    // for strcpy

struct Employee
{
   char name[30];
   int id;
   double salary;
};

void main()
{
   struct Employee q, *p;
   p = &q;
   strcpy(p -> name,"Jones, M.");
   p -> id = 17;
   p -> salary = 7777.11;
   cout << p -> name << endl;
   cout << p -> id << endl;
   cout << p -> salary << endl;
}
```

Chapter 3

String Manipulations

3.1 Introduction

C and and older C++ compilers do not support a separate string data type, but they do provide two slightly different approaches to defining strings. One is to use a character array. The other is to use a character pointer. In this section we give a number of programs for string manipulations.

Header files provide function prototype declarations for library functions. The header file

`<string.h>`

declares several string-manipulation and memory-manipulation routines. The most common are:`strlenstrcpystrcmpstrcat strncmpstrncat`

`strlen()`   `strcpy()`   `strcmp()`   `strcat()`   `strncmp()`   `strncat()`

The function `strlen()` finds the length of a string. We also give an implementation of this function. Furthermore, we show that the string length can be found recursively. The function `strcpy()` copies one string into another. The function `strcmp()` compares one string to another. We also give an implementation of this function. The function `strcat()` appends one string to another. The function `strncmp()` compares a portion of one string to a portion of another. The function `strncat()` appends a portion of one string to another.

Remark. We must always make sure that enough memory is allocated (using the operator `new`) for the string. The `new` and `delete` operators offer dynamic storage allocation and deallocation, similar but superior to the standard library functions `malloc` and `free`.

3.2 Digit Conversion to Character and String

The digits are 0, 1, 2, ..., 9. Given a digit, say 5, we want to convert it into a character '5' and a string "5". The following program does this conversion using the American Standard Code for Information Interchange (ASCII). The character code for the digit 0 is 48. Character codes for digits 0 through 9 are contiguous. Thus the character code for the digit 9 is 57. The function `map1` converts the digit into the corresponding character. The function `map2` converts the character into the corresponding string.

```
// charnum.cpp

#include <iostream.h>
#include <string.h>    // for strcpy, strlen
#include <stdlib.h>

char map1(int n)
{
   char c;
   c = 48 + n;    // ASCII 48 = 0, 49 = 1, ... , 57 = 9
   return c;
}

char* map2(char c)
{
   char* str = new char[2];
   str[0] = c;    str[1] = '\0';
   return str;
}

void main()
{
   int d;  char ch;
   cout << "\nEnter a digit from 0 to 9: ";  cin >> d;
   ch = map1(d);
   cout << "Digit: " << d << "\n" << "Character: " << ch << endl;
   char* s = map2(ch);
   char ms[2];
   strcpy(ms,s);
   delete [] s;
   cout << "String: " << ms;
}
```

3.3 Character in String

The following program finds the character at a given position in a string.

```
// charat.cpp

#include <iostream.h>
#include <string.h>
#include <stdlib.h>

char charAt(char* s,int pos)
{
   return s[pos];
}

int main()
{
   char* st = NULL;
   st = new char[15];
   strcpy(st,"microprocessor");

   int pos = 7;
   if(pos > strlen(st))
   {
   cout << "position out of range";
   exit(0);
   }

   char c = charAt(st,pos);
   cout << "c = " << c << endl;

   delete st;

   return 0;
}
```

3.4 Append a Character to a String

Given a string, say

"Willi"

We want to append a character, say '-' to this string. The output will be the string

"Willi-"

The function `char *Strcat()` in the program performs such a task. The function `strlen` determines the length of the string.

```
// appchar.cpp

#include <iostream.h>
#include <string.h>  for strlen, strcpy

char* Strcat(char* rec,char ch)
{
   size_t len = strlen(rec);
   rec[len] = ch;
   rec[len+1] = '\0';
   return rec;
}

int main()
{
   char name[10];
   strcpy(name,"Willi-");

   char c = '-';
   char *result = Strcat(name,c);
   cout << result << endl;

   return 0;
}
```

3.5 Searching for a Character in a Given String

The following program searches a string for a given character. The output is the first position of the character in the string. For example, if we consider the string `"address"` and ask for the character `'d'` we obtain the position 1. Notice that the counting starts from zero. What happens if the character is not in the string ? The function `strchr` returns a pointer to the first occurence of a particular character in a string. When we set

```
ptr = strchr(string,'d');
```

then the pointer variable `ptr` will be assigned the address of the first occurrence of the character `'d'` in the string.

```
// searchst.cpp

#include <iostream.h>  // for cin, cout
#include <string.h>    // for strchr

int main()
{
   char ch;
   char str[81];
   char* ptr;

   cout << "\n" << "Enter the string to be searched: ";
   cin.get(str,79);
   cout << "Enter character to be searched: ";
   cin >> ch;
   cout << endl;
   ptr = strchr(str,ch);     // returns pointer to char
   cout << "The character is at position: " << ptr - str;
   cout << "\n" << "Counting starts from zero";

   return 0;
}
```

3.6 Delete a Character from a String

Given a string, say

"warter"

We would like to delete the character r in "warter" in order to obtain the string "water". The following program deletes in a given string the character at a given position. Notice that the position is counted from zero. Thus the first 'r' in the string "warter" is at position 2. The function

```
char* strcpy(char *dest, const char *src);
```

copies string src to dest, stopping after the terminating null character has been moved.

```
// delchar.cpp

#include <iostream.h>
#include <string.h>      // for strcpy

void chardel(char str[], int n);  // prototype

int main()
{
   char str[81];
   int pos;
   cout << endl << "enter string: ";
   cin.get(str,79);
   cout << endl;
   cout << "position: ";
   cin >> pos;
   chardel(str,pos);  // delete character at position pos
   cout << str;

   return 0;
}

void chardel(char str[],int n)
{
   strcpy(&str[n],&str[n+1]);     // & address operator
}
```

3.7 Replacement of a Character by a Character

The function replace in the following program replaces at all occurances of a character in a string by another character.

```
// replace.cpp

#include <iostream.h>
#include <string.h>      // for strcpy

void replace(char* s,char x,char y)
{
   while(*s != '\0')
   {
   if(*s == x) *s = y;
   s++;
   }
}

int main(void)
{
   char* str = NULL;
   str = new char [7];
   strcpy(str,"tuliot");

   replace(str,'t','s');
   cout << "str = " << str;

   delete [] str;

   return 0;
}
```

3.8 Swapping Characters in a String

Consider a string, for example "susanna". We want to swap two characters in the string. Notice that the first character is at position zero. The total length of the string "susanna" is 7. At position 6 is the character 'a'. The function swap() in the two programs takes the string and the two integers of the positions at which the characters should be swapped.

```
// swapch1.cpp

#include <iostream.h>

void swap(char *s,int i,int j)
{
   char temp;
   temp = *(s+i);   *(s+i) = *(s+j);   *(s+j) = temp;
}

void main()
{
   char *str = "susanna";
   swap(str,2,3);
   cout << " string = " << str;  // output: suasnna
}
```

```
// swapch2.cpp

#include <iostream.h>

void swap(char *s,int i,int j)
{
   char temp;
   temp = s[i];   s[i] = s[j];   s[j] = temp;
}

void main()
{
   char *str = "susanna";
   swap(str,0,6);
   cout << " string = " << str;  // output ausanns
}
```

3.9 First n Characters from a String

Consider the string "abcdefghi". We want to extract the first n characters. Extracting the first four characters gives the output "abcd". The following two programs extract the first n characters of a given string.

```
// extrac1.cpp

#include <iostream.h>

int main()
{
   const int N = 10;
   char str[N];
   int pos, number;
   cout << "enter a string: ";
   cin.get(str,N,'\0');
   cout << "number of characters to extract: ";
   cin >> number;
   for(pos=0; pos < number; pos++)
   cout << str[pos];
   cout << endl;
   return 0;
}
```

```
// extrac2.cpp

#include <iostream.h>

int main()
{
   const int N = 10;
   char strg[N];
   char *str = strg;
   int pos, number;
   cout << "enter a string of " << N-1 << " characters: ";
   cin.get(strg,N,'\0');
   cout << "number of characters to extract: ";
   cin >> number;
   for(pos=0; pos < number; pos++)
   cout << *str++;

   return 0;
}
```

3.10 Substring in String

The following program displays the substring of a given string, where the first position of the first character of the substring and the position of the last character of the substring are given.

```
// substring.cpp

#include <iostream.h>
#include <string.h>

void substring(char* s,char* subs,int begin,int end)
{
   int i;
   for(i=0; i<(end-begin);i++)
   {
   subs[i] = s[i+begin];
   }
}

int main(void)
{
   int length = 15;
   int begin = 2;
   int end = 7;

   char* st = NULL;
   st = new char[length];
   strcpy(st,"Microprocessor");

   char* subst = NULL;
   subst = new char[end-begin];

   substring(st,subst,begin,end);

   cout << "subst = " << subst;

   delete st;

   delete subst;

   return 0;
}
```

3.11 Assignment and the Function strcpy

For the assignment of two strings the assignment operator = or the function `strcpy` can be used. Consider first program `kamu1.cpp`. The addresses of `s` and `t` are different because these are the addresses for the pointers. The actual pointer values (lines 3 and 7 of the main function) are the same owing to the pointer assignment statement, i.e. the two pointers point to the same memory area. The pointer `t` has no memory of its own reserved. Releasing the memory of `s` would result in an illegal pointer value for `t`. Consider now program `kamu2.cpp`. In this case `s` and `t` are different pointers and hence they have their own addresses. However, now `t` has its own memory area. Therefore this time the values of the pointer variables differ. The actual string is copied via the `strcpy` function from one memory area to the other.

```
// kamu1.cpp
#include <iostream.h>

void main()
{
   char* s = "kamu";
   cout << "&s = " << &s << endl;                   // 0x8fb9fff4
   cout << "(void*)s = " << (void*)s << endl; // 0x8fb900aa
   char* t = s;
   cout << "&t = " << &t << endl;                   // 0x8fb9fff2
   cout << "(void*t) = " << (void*)t << endl; // 0x8fb900aa
   cout << "t = " << t << endl;                     // kamu
}
```

```
// kamu2.cpp
#include <iostream.h>
#include <string.h>

void main()
{
   char* s = "kamu";
   cout << "&s = " << &s << endl;                   //0x8fbcfff4
   cout << "(void*)s = " << (void*)s << endl; //0x8fbc00aa
   char* t;   t = new char[5];
   strcpy(t,s);
   cout << "&t = " << &t << endl;                   //0x8fbcfff2
   cout << "(void*)t = " << (void*)t << endl; //0x8fbc0d14
   cout << "t = " << t << endl;                     //kamu
}
```

3.12 Permutations of a Word

The following program generates all possible *permutations* of a given word. For example, if the input is

```
ute
```

the output will be

```
uet   tue   teu   eut   etu
```

If a word has n letters (characters) then the number of permutation is

$$n! := 1 \cdot 2 \cdot \ldots \cdot n\,.$$

The function `permute` prints out all the permutations of whatever word is given it. The function `rotate` rotates all the characters in the string `word`, starting at position `startrot`. The function `permute` uses the function `rotate`.

```
// anagram.cpp

#include <iostream.h>
#include <string.h>     // for strlen

void permute(int);    // prototype
void rotate(int);     // prototype

     // global declaration
char word[20];
int leng;         // length of string (word);

int main()
{
   cout << " enter word: ";
   cin >> word;
   leng = strlen(word);
   permute(0);

   return 0;
}
```

```
void permute(int startperm)
{
   int j;

   if(leng - startperm < 2)
   return;

   for(j=startperm;j<leng-1;j++)
   {
   permute(startperm+1);
   rotate(startperm);
   cout << "\n" << word;
   }
   permute(startperm+1);
   rotate(startperm);
}

void rotate(int startrot)
{
   int j;
   char ch;
   ch = word[startrot];
   for(j=startrot;j<leng-1;j++)
   word[j] = word[j+1];

   word[leng-1] = ch;
}
```

3.13 String within String

The following two programs find the last occurrence of a string s2 in a string s1. If there is no match then there is no output. As an example we consider the string "willilo". The string we search for is "il". Thus the output will be "ilo". The function strstr scans a string for the occurrence of a given substring. The condition if(!strstr(s1,s2)) checks whether there is a match or not.

```
// laststr1.cpp

#include <iostream.h>
#include <string.h>      // for strstr()

// strrstr() finds last occurrence of s2 in s1
char *strrstr(char *s1,char *s2)
{
   char *old = NULL;
   if(!strstr(s1,s2))
   return NULL;          // if no match just return
   while(s1)  {
     old = s1;
     s1 = strstr(s1+1,s2);
   }
   return old;
}

void main()
{
   char *s1 = "willilo";
   char *s2 = "il";
   char *s3;
   s3 = strrstr(s1,s2);
   cout << s1 << "\n";  // output: willilo
   cout << s2 << "\n";  // output: il
   cout << s3 << "\n";  // output: ilo

   char *r1 = "nation";
   char *r2 = "i";
   char *r3;
   r3 = strrstr(r1,r2);
   cout << r3 << "\n";  // output: ion
}
```

```
// laststr2.cpp

#include <iostream.h>
#include <string.h>   // for strstr()

// strrstr() finds last occurrence of s2 in s1
char *strrstr(char *s1,char *s2)
{
   char *s, *t, *u;
   for(s=s1;*s;s++)
   if(*s2=='\0') return s;  // '\0' can also be written as NULL
   while(s-- > s1)
   {
   for(t=s,u=s2;*u && *t == *u;t++, u++);
   if(!*u) return s;
   }
   return NULL;
}

void main()
{
   cout << endl;
   char *s1 = "willilo";
   char *s2 = "il";
   char *s3 = NULL;
   s3 = new char[4];
   strcpy(s3,strrstr(s1,s2));
   cout << s1 << "\n";  // output: willilo
   cout << s2 << "\n";  // output: il
   cout << s3 << "\n";  // output: ilo
   delete s3;

   char *r1 = "nation";
   char *r2 = "i";
   char *r3 = NULL;
   r3 = new char[4];
   strcpy(r3,strrstr(r1,r2));
   cout << r3 << "\n";  // output: ion
   delete r3;
}
```

3.14 Comparing Strings

In the following two programs we compare strings. The library function `strcmp` which is included in the header file `<string.h>` was described and applied in section 2.10. Here we give two implementations.

```
// comstr1.cpp

#include <iostream.h>

int strcmp(char *s,char *t)
{
   int i;
   for(i=0;s[i]==t[i];i++)
      if(s[i] == '\0')
      return 0;
   return s[i]-t[i];
}

void main()
{
   char *str1 = "Willi";
   char *str2 = "Hans";
   int result;

   result = strcmp(str1,str2);
   cout << "result = " << result << "\n";
   if(result < 0)
   cout << "str1 < str2" << "\n";
   if(result == 0)
   cout << "str1 = str2" << "\n";
   if(result > 0)
   cout << "str1 > str2" << "\n";
}
```

```
// comstr2.cpp

#include <iostream.h>

int strcmp(char *s,char *t)
{
   int i;
   for(;*s==*t; s++, t++)
   if(*s == '\0')
   return 0;
   return *s - *t;
}

int main()
{
   char *str1 = "Willi";
   char *str2 = "Hans";
   int result;

   result = strcmp(str1,str2);
   cout << "result = " << result << "\n";
   if(result < 0)
   cout << "str1 < str2" << "\n";
   if(result == 0)
   cout << "str1 = str2" << "\n";
   if(result > 0)
   cout << "str1 > str2" << "\n";

   return 0;
}
```

3.15 Length of a String

In the following program we determine the length of a given string. For example, the length of the string `"willi"` is 5. The length of the string `"willi-hans"` is 10. Two functions `length1` and `length2` are implemented to find the length of the string. C and C++ have a built in function `strlen` to find the length of a string (included in header file `<string.h>`). Recall that by convention all character arrays are terminated with a `'\0'`. The pointer version is more efficient (this is almost always true). Even more compact is `for (m=0; *s++ != '\0'; m++)`. The `*s` returns a character. The `++` increments the pointer so we get the next character next time around.

```
// length.cpp

#include <iostream.h>

int length1(char s[])
{
  int n;
  for(n=0;s[n] != '\0'; )
  n++;
  return n;
}

int length2(char *s)
{
  int m;
  for(m=0;*s != '\0';s++)
  m++;
  return m;
}

int main()
{
  cout << "\n" << length1("Willi") << "\n";
  cout << length2("Hans") << "\n";
  char *u = "Dynamics";
  cout << length1(u) << "\n";
  char v[] = "Equations";
  cout << length2(v);
  return 0;
}
```

3.16 Permutations of Characters in a String

The program prints every permutation of the characters in the string we supply on the command line. After generating the execute file `permstr.exe` at the command prompt we type

```
permstr string
```

This routine immediately recurses, passing itself a copy of the string, until it has recursed to a depth equal to the length of the string. At that point it loops, printing the string and shifting left the character located in the array offset by the recursion level. When the character being shifted reaches the beginning of the string, it drops back to the previous level of recursion. The function `permute` is the front-end to the routine that does the real work. It allocates the memory that the routine will need, calls the routine, and cleans up. The following program uses only C commands, except that the function `malloc` is replaced by the `new` operator and `free` by `delete`.

```
// permstr.cpp

#include <stdio.h>
#include <stdlib.h>     // for exit
#include <string.h>     // for strcpy, strlen

void permute_engine(char *string,int level,char **tmparea)
{
   char *copy = tmparea[level];
   char tmpc;
   int index = level;
   int printing = string[index + 1] == '\0';

   strcpy(copy, string);
   do
   {
   if(printing)
   puts(copy);
   else permute_engine(copy,level+1,tmparea);
   if(index != 0)
   {
   tmpc = copy[index];
   copy[index] = copy[index-1];
   copy[index-1] = tmpc;
   }
   } while(index-- > 0);
```

```
}

void permute(char *str)
{
   int x;
   int len = strlen(str) + 1;
   char **tmparea = NULL;
   if((tmparea = new char* [len*sizeof(*tmparea)]) == NULL)
   {
   fputs("Error: Out of memory.\n", stderr);
   exit(1);
   }
   for(x=len-2; x >= 0; x--)
   {
   if((tmparea[x] = new char[len]) == NULL)
   {
   fputs("Error: Out of memory.\n", stderr);
   exit(1);
   }
   }
   permute_engine(str,0,tmparea);

   for(x=len-2; x >= 0; x--)
   delete(tmparea[x]);
   delete tmparea;
}

int main(int argc, char *argv[])
{
   if(argc != 2)
   {
   fputs("Usage: permstr string\n",stderr);
   return 1;
   }
   permute(argv[1]);

   return 0;
}
```

3.17 Swapping Parts of a String

Given a string, for example

    "willihanssteeb"

We want to swap parts around. If `length` is the length of the string (in the given example 14) and `partition` is set to 2, then the output is

    "llihanssteebwi"

The function `memcpy(void *dest, const void *src, size_t n)` copies a block of n bytes from `src` to `dest`.

```
// partswap.cpp

#include <iostream.h>
#include <string.h>

void swap_parts(char *array,int length,int partition)
{
   char *small_part;
   int small_length = length-partition;
   register int j;

   small_part = new char [small_length+1];
   memcpy(small_part,array+partition,small_length);

   for(j=length-1; j>=small_length; j--)
   array[j] = array[j-small_length];

   memcpy(array,small_part,small_length);
   delete small_part;
}

void main()
{
  char *s = "willihanssteeb";
  int m = strlen(s);
  int n = 2;
  swap_parts(s,m,n);
  cout << "s = " << s << "\n";
}
```

3.18 String Handling using Recursion

A string consists of contiguous characters in memory, ending with the NULL character '\0'. Conceptually, we can think of a string as either the NULL string, consisting of just the NULL character, or as a character followed by a string. This definition of a string describes it as a recursive data structure. Thus we can use this to code basic string handling functions recursively.

The first program leng.cpp determines the length of a string recursively.

```
// leng.cpp
// recursive string length

#include <iostream.h>

int r_strlen(char *s)
{
   if(*s == '\0')
      return 0;
   else
      return(1 + r_strlen(s + 1));
}

int main()
{
   char *st = "union";
   int res = r_strlen(st);
   cout << "length is: " << res << endl;

   return 0;
}
```

A recursive version for the standard library function strncmp is given in the following. It lexicographically compares at most the first n characters of two strings. The function strncmp compares a portion of one string to a portion of another. The function strncmp returns an integer value based on the result of comparing string s1 (or part of it) to string s2 (or part of it).

```
// rec.cpp
// recursive string n compare

#include <iostream.h>

int r_strncmp(char *s1,char *s2,int n)
{
   if(*s1 != *s2 || *s1 == '\0' || n == 1)
      return (*s1 - *s2);
   else
      return (r_strncmp(++s1,++s2,--n));
}

int main()
{
   char *st1 = "unipn";
   char *st2 = "union";
   int res1 = r_strncmp(st1,st2,3);

   cout << "res1 = " << res1 << endl;  // output 0

   int res2 = r_strncmp(st1,st2,4);
   cout << "res2 = " << res2 << endl;  // output 1

   int res3 = r_strncmp(st2,st1,4);
   cout << "res3 = " << res3 << endl;  // output -1

   return 0;
}
```

3.19 Concatenating Strings

In the following program we concatenate strings together. First we copy "hello" into buffer using strcpy. Then we concatenate the strings "Nela ", "how ", "are ", "you ?" to it using the function strcat. Both functions strcpy and strcat are included in the header file <string.h>.

```
// concat1.cpp

#include <iostream.h>  // for cout
#include <string.h>    // for strcpy(), strcat()

int main()
{
   cout << endl;
   char buffer[100];
   strcpy(buffer,"Hello ");
   strcat(buffer,"Nela, ");
   strcat(buffer,"how ");
   strcat(buffer,"are ");
   strcat(buffer,"you ?");

   cout << endl << buffer << endl;
   cout << "length of string = ";
   cout << strlen(buffer);
   cout << endl;
   cout << "buffer[10] = " << buffer[10] << endl;

   return 0;
}
```

The output is

```
Hello Nela, how are you ?
```

```
// concat2.cpp

#include <iostream.h>
#include <string.h>    // for strcpy(), strcat(), strlen()

int main()
{
   cout << endl;
   char* string1 = "Willi-";
   int leng1 = strlen(string1);
   char* string2 = "Hans";
   int leng2 = strlen(string2);

   char* string3 = NULL;
   string3 = new char[leng1+leng2+1];

   strcat(string3,string1);
   strcat(string3,string2);

   cout << "resulting string: " << string3 << endl;
   int leng3 = strlen(string3);
   cout << "length of string: " << leng3 << endl;

   delete string3;

   return 0;
}
```

3.20 Replacing a Substring in a String

Given a string, say "Willi". We want to replace the substring "ll" by "pp" in oder to obtain "Wippi". The function sub in the following program replaces a substring in a given string by another string.

The following string functions are used: strlen, strcpy, strncat, strcat. The function strncat appends a portion of one string to another string. The function strstr scans a string for the occurrence of a given substring.

```
// insert.cpp

#include <iostream.h>
#include <string.h> // for strlen(), strcpy(), strstr(),
// strncat(), strcat()

void sub(char oldstr[],char repostr[],char repnstr[])
{
   int rlen = strlen(repostr);
   char *pstr;
   pstr = oldstr;
   char p2str[255], wstr[255];

   strcpy(wstr,oldstr);
   strcpy(p2str, "");

   while((pstr = strstr(pstr,repostr)) != NULL)
   {
   strncat(p2str,wstr,strlen(wstr)-strlen(pstr));
   strcat(p2str,repnstr);
   pstr = pstr + rlen;
   strcpy(wstr,pstr);
   }

   strcat(p2str,wstr);
   strcpy(oldstr,p2str);
   return;
}
```

```
int main()
{
   char oldstr[255], repostr[255], repnstr[255];

   cout << "Enter string = ";
   cin >> oldstr;

   cout << "Enter substring to be replaced = ";
   cin >> repostr;

   cout << "Enter replacing substring = ";
   cin >> repnstr;

   // call function sub
   sub(oldstr,repostr,repnstr);

   cout << "replaced string : " << oldstr << endl;

   return 0;
}
```

3.21 Reversing the Characters in a String

The following program reverses the characters in a string. For example if we have the word "united nation" we obtain "noitan detinu". The function reverse first finds the end of the given string. It then backs off one position from '0' so that the first character will not become an end of string marker.

```
// invert.cpp

#include <iostream.h>

void reverse(char s[]);   // prototype

int main()
{
   const int MAX = 100;
   char word[MAX] = "united nation";
   reverse(word);
   cout << word << "\n";
   return 0;
}

void reverse(char s[])
{
   int i, j;
   char t;
   i = 0;
   while(s[i] != '\0')  // find the end of string
   ++i;
   --i;                 // back off from '\0'
   if(s[i] == '\n')
   --i;
   j = 0;
   while(j < i)
   {
   t = s[j];
   s[j] = s[i];     // swap the characters
   s[i] = t;
   --i;
   ++j;
   }
}
```

The following program reverses the characters between s[j] and s[k] recursively. This means that within the function reverse() the function invokes itself.

```
// rev.cpp
// reverse the characters between s[j] and s[k] recursively

#include <iostream.h>

void  reverse(char *s,int j,int k);
void  swap(char *,char *);

int main()
{
   cout << endl;
   char  phrase[] = "by the sea, by the beautiful sea";
   // the characters from position 3 to position 17
   // have been reversed, counting from 0
   reverse(phrase,3,17);
   cout << phrase;
   return 0;
}

void reverse(char *s,int j,int k)
{
  if(j < k)
  {
  swap(&s[j], &s[k]);
  reverse(s, ++j, --k);
  }
}

void swap(char *p,char *q)
{
   char tmp;
   tmp = *p;
   *p = *q;
   *q = tmp;
}
// output:
// by eht yb ,aes eht beautiful sea
```

3.22 Boyer and Moore's Algorithm

A common task in string manipulation is to search a string (or file) for the occurrences of a particular substring (or piece of text). What is needed is a function which could take a string such as

```
Isn't it funny how bees fly
```

and a substring such as `funny` and tell us that the substring occurs in the string at character position 9 (recall that the first character position is numbered zero). The straightforward approach is to scan the string for an `f` and when one is found the characters after the `f` are checked to see whether they match. While this approach is simple, it is not the best option. It involves checking every character of the string, which can be very slow.

A better approach is given by Boyer and Moore. In their method a table is built from the substring indicating the position of the last occurrence of each character within it. For the string `funny` this would hold the information that

```
f at 0, u at 1, n at 2, n at 3, y at 4.
```

The algorithm then involves searching in the string for the last character of the susbstring as follows: the substring contains five characters, numbered zero to four. Obviously, we must not check the string before character number 4, which is a `t`. The table is checked and reveals that there is no blank in the substring and therefore we can skip five characters to the `f`. This is in the table and is at position zero in the substring . Thus four characters can be skipped to the `y`. This is the character being searched for and so a comparison is made at this point between the whole substring and the appropiate part of the string. This operation finds a match.

```
// boyer.cpp

#include <iostream.h>
#include <stdio.h>
#include <string.h>
#include <limits.h>

const int charsetsize = 128;

int substringposition(char *string, int stringleng, char *substr,
int substrleng, int contentlist[])
{
   int result, index, position, firstchar;
   result = -1;
   char finalchar, currentchar;
   finalchar = substr[substrleng-1];
   char *stringptr, *substrptr;
```

```
    if(stringleng >= substrleng)
    {
      if(substrleng == 1)
      {
      for(index = 0;index < stringleng && result == -1;++index)
      if(finalchar == string[index])
      result = index;
      }
      else
      {
      index = substrleng-1;
      do
      {
      if((position = contentlist[currentchar = string[index]]) == -1)
      index += substrleng;
      else
      if(currentchar == finalchar)
      {
      firstchar = index - substrleng + 1;
      if(strncmp(&string[firstchar],substr,substrleng) == 0)
      result = firstchar;

      if(result == -1)
      ++index;
      }
      else
      index += (substrleng - position - 1);
      }
      while(index < stringleng && result == -1);
      }
      }
      return result;
}

void makecontentlist(char *substr, int contentlist[])
{
   int i;
   for(i=0; i<= charsetsize-1; ++i)
   contentlist[i] = -1;

   for(i=0; *substr != '\0'; ++i, ++substr)
   contentlist[*substr] = i;
}
```

```
int main()
{
   const int lineleng = 132;
   const int nameleng = 12;

   int line = 1;
   int searchstringleng;
   int contentlist[charsetsize];

   FILE *InStream;
   char currentline[lineleng], filename[nameleng],
   searchstring[lineleng];

   cout << endl << "enter name of file to be searched: ";
   gets(filename);
   InStream = fopen(filename,"r");
   cout << endl;
   cout << "enter the substring to be searched: ";
   gets(searchstring);

   makecontentlist(searchstring,contentlist);
   searchstringleng = strlen(searchstring);
   while(fgets(currentline,lineleng,InStream) != NULL)
   {
   if(substringposition(currentline,strlen(currentline),
   searchstring,searchstringleng,contentlist) != -1)
   printf("line: %d text: %s",line,currentline);
   ++line;
   }

   return 0;
}
```

3.23 Subsets of a Given Set of Strings

Given a set S of n strings, say

```
"balloon", "tree", "fish", "dog", "cat"
```

We want to generate all possible m item subsets of S. This means display all subsets of m elements which can be found in a set of n elements ($m \leq n$). The function `showset` works recursively. It builds each qualifying subset in a single allocated buffer. Notice that the number of subsets of a finite set with n elements is 2^n, which include the set itself and the empty set.

```
// subset.cpp

#include <iostream.h>

static char *set[] = { "balloon", "tree", "fish", "dog", "cat" };

void showset(char *set[],int n,int m)
{
    int i, j;
    static int sp = 0;
    static char **subset;
    if(!sp)
    subset = new char* [m*sizeof(char *)]; //allocate memory
    for(i=0; i < n;) {
    subset[sp++] = set[i++];     // add element to subset
    if(sp == m) {               // correct size subset - print
    for(j=0; j < sp; ++j)
    cout << subset[j] << " ";
    cout << "\n";
    }
    else                         // expand subset to the right
    showset(&set[i],n-i,m);
    --sp; }
    if(!sp)
    delete subset;    // release allocated memory
}

int main()
{
    showset(set,sizeof(set)/sizeof(set[0]),2);
    return 0;
}
```

3.24 Counting Characters and Digits in a Phrase

The number of characters and digits in a phrase typed in are counted. For example if the phrase

```
The longitude is 26 30E
```

is entered, the output of the program is:

```
Number of characters are: 30

Number of digits are: 4
```

```
// digitcnt.cpp

#include <iostream.h>
#include <conio.h>     // for getche()

int main()
{
   int nc = 0;
   int nd = 0;
   char ch;

   cout << " enter a phrase: ";

   while((ch = getche()) != '\r')   // until [Return] typed
   {
   nc++;                            // count characters
   if(ch > 47 && ch < 58)           // if ch is digit, ASCII-code
   nd++;                            // count digits
   }
   cout << "\n" << " number of chacters are: " << nc;
   cout << "\n" << " number of digits are: " << nd;

   return 0;
}
```

3.25 Counting Characters and Words in a Phrase

The number of characters and words in a phrase typed in are counted. For example if the phrase

```
The longitude is 26 30E
```

is entered, the output of the program is:

```
Number of characters are: 30

Number of words are: 5
```

```
// wordcnt.cpp

#include <iostream.h>
#include <conio.h>      // for getche()

int main()
{
   int nc = 0;     // number of characters
   int nw = 0;     // number of words
   char ch;

   cout << " enter a phrase: ";

   while((ch=getche()) != '\r')    // read character and
   {                              // quit loop on [Return]
   nc++;                          // count character
   if(ch == ' ')                  // space ?
   nw++;                          // count word
   }
   cout << "\n" << "number of characters are: " << nc;
   cout << "\n" << "number of words are: " << nw+1 << "\n";

   return 0;
}
```

```
// word1.cpp
// press ESC to exit

#include <iostream.h>
#include <conio.h>

int main()
{
   int word = 0;
   char ch = 'a';
   int count = 0;
   cout << "Enter sentence: ";
   do
   {
   ch = getche();
   if(ch == ' ' || ch == '\33')
   {
   if(word)
   {
   count++;
   word = 0;
   }
   }
   else
   if(!word)
   word = 1;
   } while(ch != '\33');  // press ESC to exit

   cout << "\n" << "Number of words: " << count << "\n";

   return 0;
}
```

3.26 Centering Text

The following program centers text. The function `sprintf` writes formatted output to a string.

```
// center.cpp

#include <stdio.h>   // printf, sprintf
#include <string.h>  // strlen

char *verse[] =
{
   "A lover's eyes will gaze an eagle blind;",
   "A lover's ear will hear the lowest sound,",
   "When the suspicious head of theft is stopp'd:",
   "Love's feeling is more soft and sensible",
   "Than are the tender horns of cockled snails:",
   "William Shakespeare",
   "Love's Labour's Lost",
   NULL
};

int main()
{
   char **ch_pp;

   for(ch_pp=verse;*ch_pp;ch_pp++)
   {
   int length;
   char format[10];
   length = 40 + strlen(*ch_pp)/2;
   sprintf(format,"%%%ds\n",length);
   printf(format,*ch_pp);
   }
   printf("\n");

   return 0;
}
```

3.27 The Function getline

The function `getline` is an inherited member function of `cin`, declared in

`iostream.h`

The function `getline` reads characters until encountering the `'\n'` character and places the resulting string in the buffer specified by the argument. The following program gives an application of this function. The nested `for` loop prints a number of `*` equal to the length of the string.

```
// line.cpp

#include <iostream.h>
#include <string.h>    // for strcmp, strlen

int main()
{
   int i;
   char data[80];

   cout << "enter string: " << endl;
   cout << "to quit enter esc" << endl;

   for(cin.getline(data,80); strcmp(data,"esc") != 0;
       cin.getline(data,80))
   {
   for(i=1; i <= strlen(data); i++)
   cout << "*";
   cout << endl;
   }

   return 0;
}
```

Chapter 4

The Class Concept

4.1 Introduction

Object-oriented programming is the most dramatic innovation in software development in the last decade. It offers a new and powerful way to cope with the complexity of programs. Using object-oriented programming we can write clearer, more reliable, more easily maintained programs. Integrating data and functions is the central idea of object-oriented programming. Of the object-oriented programming languages, C++ is by far the most popular. C++ is a superset of C. All programs in C should also run under C++.

C++ not only corrects most of C's deficiencies, it also introduces many completely new features that were designed into the language to provide support for data abstraction and object-oriented programming. Here are some of the more prominent new features:

- *classes*, the basic language construct for creating programmer defined data types that we call abstract data types.

- *member variables*, which describe the data in abstract data types, and *member functions* (also called *methods*), which define the permissible operations on the data type.

- *operator overloading*, which lets us give additional meaning to most operators so that we can use them with our own data types, thereby making those data types easier to use, and *function name overloading*, similar to operator overloading, which lets us reduce the need for unusual function names, making code easier to read.

- *programmer-controlled automatic type conversion*, which lets us blend our own types with others and with the basic data types provided by the C++ language.

- *derived classes*, which inherit member variables and member functions from their base classes, and can be differentiated from their base classes by adding other member variables and member functions.
- *virtual functions*, which let a derived class redefine member functions inherited from a base class. We can then write very general programs that are oblivious to the specific classes of the objects they manipulate. Through *dynamic binding*, the run time system will choose the function appropriate to a particular class.
- *function templates* provide an "algorithm" for the automatic generation of particular instances of that function varying by type.
- *class templates* provide for the automatic generation of class instances bound to a particular type. Unlike the macro facility, it is an integral part of the C++ language.
- *standard template library* provides a number of useful classes.
- built in *string class* for newer C++ compilers
- boolean data type *bool* added

We show in succeeding programs how to apply these new tools provided by C++ to create data types that are suitable for particular applications, and then to combine those abstract data types into object-oriented programs.

C++ is compatible with C, and because C++ uses the fundamental data types – *int, float* and so forth – in the conventional way, it does not force the object-oriented paradigm on programmers. We can continue to write conventional procedural programs just as we did in C, and still take the advantage of many of C++'s new features.

A complete list of operators which can be overloaded can be found in the C++ reference guides.

4.2 What is a Class ?

A `class` defines a data type. In a computer science sense, a type consists of two things: a set of values and a set of operations which operate on those values. Thus `int` all by itself is not a true type until we add operations like add two integers or multiply two integers, etc. In exactly the same way, a `class` provides a set of usually `public` operations, and a set of usually `private` data bits representing the abstract values that instances of the type can have. From a C language perspective, a `class` is a `struct` whose members default to `private`.

C++ enables us to declare a `class` that encapsulates data members and member functions. The member functions alter the state of the `class` instances (objects) by changing the various values of one or more data members. The general syntax for declaring a base `class` (one that is not derived from any other class) is:

```
class className
{
    private:
 <private data members>
 <private member functions>

    protected:
 <protected data members>
 <protected member functions>

    public:
 <public data members>
 <public member functions>
};
```

The preceding syntax shows that the declaration involves the `class` keyword. The definition of a member function can also be placed inside the class body. Member functions larger than one or two lines are best defined outside the `class` body. C++ classes offer three levels of visibility for the various members (that is, both data members and member functions).

An *object* is a region of storage with associated semantic. After the declaration

```
int i;
```

we say that `i` is an object of type `int`. In C++/OOP, an object is usually used to mean an instance of a `class`. Thus a `class` defines the behaviour of possibly many objects (instances).

4.3 Constructors and Destructors

There are several special member functions that determine how the objects of a class are created, initialized, copied, and destroyed. *Constructors* and *destructors* are the most important of these. Generally, an object must be initialized before use. Global variables are automatically initialized to 0, but this value often is not acceptable. In C the function that declares the object assigns a value to each of its members. Values are assigned either individually or all at once in an initialization statement. In C++, however, this process is inadequate. To obtain the maximum benefit from encapsulation, most of the data members of a C++ class should be private. The function that declares an object of such a class does not have access to assign initial values to its private members. C also permits objects to be initialized by defining a function and then invoking the function after the object is declared but before it is used. A similar approach can be used in C++ since initialization can be a member function with access to the private members. To facilitate this process, C++ allows the programmer to define a special method known as a constructor. A constructor is a special member function that is automatically invoked whenever an object to that type is created.

The part about establishing coherence is the job of a constructor. Constructors are like init functions; they build a valid object. The constructor turns a pile of incoherent arbitrary bits into a living object. Minimally it initializes any internally used fields that are needed, but it may also allocate resources (memory, files, semaphores, sockets, ...).

A constructor is like a factory: it builds objects from dust. The default constructor for `class X` is one that takes no arguments; it has the form

```
X::X()
```

If no user-defined constructors exist for a `class`, C++ generates a default constructor. On a declaration such as

```
X x;
```

the default constructor creates the object `x`.

Destructors are used to release any resources allocated by the object's constructor. For example: a Lock class might lock a semaphore, and the destructor will release that semaphore. The usual resource being acquired in a constructor (and subsequently released in a destructor) is dynamically allocated memory.

In C++, constructors and destructors work automatically to guarantee the appropriate creation and removal of class instances. C++ has the following features and rules regarding constructors:

1. The name of the constructor must be identical to the name of its class.
2. We must not include any return type, not even void.
3. A class can have any number of constructors, including none. In the latter case, the compiler automatically creates one for us.
4. The default constructor is the one that has neither parameters nor a parameter list where all the parameters use default arguments.
5. The declaration of a class instance (which includes function parameters and local instances) involves a constructor. Which constructor is called ? The answer depends on how many constructors we have declared for the class and how we declared the class instance.

1. The `private` section: only member functions of the class can access the private members. The class instances are denied access to private members.
2. The `protected` section: only member functions of the class and its descendant classes can access protected members. The class instances are denied access to protected members.
3. The `public` section: specifies members that are visible to the member functions of the `class`, `class` instances, member functions of descendant classes, and their instances.

The following rules apply to the various sections:

1. Class sections can appear in any order.
2. Class sections may appear more than once.
3. If no class section is specified, the C++ compiler treats the members as protected.
4. Avoid placing data members in the public section, unless such a declaration significantly simplifies your design. Data members are typically placed in the protected section to allow their access by member functions of descendant classes.
5. Use member functions to set and query the values of data members. The members set to the data members assist in performing validation and updating other data members, if necessary.

After we declare a class, we can use the class name as a type identifier to declare class instances. The syntax resembles declaring variables.

Constructors always carry the same name as the object itself and have no return type (not even `void`). A constructor for `class Address` might be declared as follows:

```
Address(int number,char* street,char* city,char* state,char* zip);
```

The arguments are passed to the constructor when the variable is created. To invoke the above constructor, an object of type Address would be declared as follows:

```
Address myAddr(105, "Rockcrest St.", "Mesquite", "TX", "75401");
```

The arguments appearing after `myAddr` serve the same purpose as an initializer list following an equal sign in C. C++ also allows the programmer to define a member function that is called when an object is destroyed. This method, called a destructor, carries the name of the class preceded with a ~. A destructor has no return type and takes no arguments. The destructor for Address is declared as follows:

```
~Address();
```

4.4 Copy Constructor

A *copy constructor* for a `class X` is a constructor that can be called to copy an object of `class X`; that is, one that can be called with a single argument of type `X`. For example,

```
X::X(const X&)                    X::X(X&,int=0)
```

are copy constructors. A copy constructor is generated only if no copy constructor is declared. A copy constructor for a `class X` may not take an argument of type `X`. For example

```
X::X(X)
```

is illegal.

Copy constructors are invoked when initializing a class object, typically when one declares with initialization by another class object:

```
X x1;
X x2 = x1;
X x3(x1);
```

Both styles of definition invoke a copy constructor: that is, a constructor that copies its arguments into a new object. The default copy constructor, which is provided automatically by the compiler for every object, performs a member-by-member copy. This is similar to what the assignment operator does. The *assignment operator* = is declared as follows

```
X& operator = (const X&);
```

The difference is that the copy constructor also creates a new object.

C++ generates a copy constructor for `class X` if one is needed and no other copy constructor has been defined in `class X`. The copy constructor that is generated by the C++ compiler allows us to safely start programming with simple data types. We need to make our own definition of the copy constructor only if our program creates aggregate, complex types such as `class`, `struct`, and arrays.

Many C++ programmers strongly recommend that we declare copy constructors, especially for classes that model dynamic data structures such as linked lists.

4.5 Operator Overloading

Operator overloading is one of the most exciting features of object-oriented programming. It can transform complex, obscure program listings into intuitively obvious ones. The term *operator overloading* refers to giving the normal C++ operators, such as

```
+,  -,  *,  /,  <=,  +=
```

additional meanings when they are applied to user defined data types. The addition

```
c = a + b;
```

only works with basic types like `int` and `float`. We cannot add for example two vectors, even when the objects of a user-defined class. However, using overloading we can make this operation legal even when a, b, and c are user defined types. Operator overloading gives us the opportunity to redefine the C++ language.

We could ask the question why can we not use

```
X operator = (X)
```

instead of

```
X& operator = (const X&)
```

Passing by value is daft: the copy constructor will be invoked to initialize `arg` with a copy of the actual argument, when all we want to do is to copy the `arg` values. The argument should be passed by reference, to avoid doing the copying twice, and what is more, it should be passed by const reference, since we are not going to change it. The operator = should return a reference to the object that has just assigned to, which then can be used in a chain of assignments.

Applications where one applies operator overloading are classes for the rational numbers, complex numbers, vectors, quaternions, and matrices. The concatenation of two string can also be rewritten with +. Next we give the list of operators which can be overloaded.

```
+   -   *   /   %
=   ==
+=   -=   *=   /=   %=
&   |   ^   ~
&=   |=   ^=   ~=
!   ,   >   <
<<   >>   >>=   =<<
<=   >=
&&   ||
++   --
->   ^*   ->()   []
```

4.6 The Keyword this

Nonstatic member functions operate on the class type object with which they are called. For example, if `x` is an object of class `X` and `f` is a member function of `X`, the function call

```
x.f()
```

operates on `x`. Similarly if `xptr` is a pointer to an `X` object, the function call

```
xptr -> f()
```

operates on `*xptr`. But how does `f` know which `x` it is operation on ? C++ provides `f` with a pointer to `x` called

```
this
```

The `this` pointer is passed as a hidden argument in all calls to nonstatic member functions.

The `this` pointer is a local variable available in the body of any nonstatic member function. The `this` pointer does not need to be declared and is rarely referred to explicitly in a function definition. However, it is used implicitly within the function for member references. If

```
x.f(y)
```

is called, for example, where `y` is a member of `X`, `this` is set to `&x` and `y` is set to `this -> y` which is equivalent to `x.y`.

We use the `this` pointer when we overload the assignment operator =.

```
// myclass.cpp

#include <iostream.h>

class X
{
   public:
      X();
     ~X();
};

X::X()
{
   cout << "this = " << this << endl;
}

X::~X()
{
   cout << "this = " << this << endl;
   cout << "destructor called" << endl;
}

int main()
{
   X x;
   cout << "&x = " << &x << endl;

   X y;
   cout << "&y = " << &y << endl;

   return 0;
}
```

The output is

```
this = 0x0012FF7C
&x = 0x0012FF7C
this = 0x0012FF78
&y = 0x0012FF78
this = 0x0012FF78
destructor called
this = 0x0012FF7C
destructor called
```

4.7 Examples

4.7.1 Example 1.

```
// class1a.cpp
// calculates the areas of geometric figures
// square, rectangle, circle, ellipse

#include <iostream.h>  // for cout cin
#include <ctype.h>     // for toupper

    // Declaration of class Geometric
class Geometric
{
   private:
      double x, y;    // private data members
      double AS();    // member function calculates area of square
      double AR();    // member function calculates area of rectangle
      double AC();    // member function calculates area of circle
      double AE();    // member function calculates area of ellipse
   public:
      Geometric();                     // default constructor
      Geometric(const Geometric&);     // copy constructor
      ~Geometric();                    // destructor
      Geometric(double);               // constructor one argument
      Geometric(double, double);       // constructor two arguments
      Geometric& operator = (const Geometric&); // overloaded =
      void menu();                     // public member function
};    // end declaration class Geometric


      // Definition (Implementation) of class Geometric
Geometric::Geometric() {  }  // :: scope resolution operator

Geometric::Geometric(const Geometric& cp)
{
   x = cp.x;  y = cp.y;
}

Geometric::~Geometric()
{
   cout << endl << "destructor called" << endl;
}

Geometric::Geometric(double x1)  { x = x1; }
```

```
Geometric::Geometric(double x1,double y1) { x = x1;  y = y1; }

Geometric& Geometric::operator = (const Geometric& arg)
{
   x = arg.x;  y = arg.y;
   return *this;
}

double Geometric::AS()  { return x*x; }

double Geometric::AR()  { return x*y; }

double Geometric::AC()
{
   double pi = 3.14159;
   return pi*x*x;     // area of circle
}

double Geometric::AE()
{
   double pi = 3.14159;
   return pi*x*y;     // area of ellipse
}

void Geometric::menu()
{
   double r, a, b;
   char cmd;

   do
   {
   cout << endl;
   cout << "Chart desired: Square Rectangle Circle Ellipse Quit";
   cout << "\nPress first letter of the chart you want: ";
   cin >> cmd;
   cmd = toupper(cmd);

   switch(cmd)
   {
   case 'S': { cout << "Enter the length: ";
   cin >> a;
   Geometric square(a);
   cout << "The area of the square is: " <<  square.AS();
   break; }
```

```
   case 'R': { cout << "Enter the length of the rectangle: ";
   cin >> a;
   cout << "Enter the width of the rectangle: ";
   cin >> b;
   Geometric rectangle(a,b);
   cout << "The area of the rectangle is: " << rectangle.AR();
   break; }

   case 'C': { cout << "Enter the radius: ";
   cin >> r;
   Geometric circle(r);
   cout << "The area of the circle is: " << circle.AC();
   break; }

   case 'E': { cout << "Enter the semi-major axis of the ellipse: ";
   cin >> a;
   cout << "Enter the semi-minor axis of the ellipse: ";
   cin >> b;
   Geometric ellipse(a,b);
   cout << "The area of the ellipse is: " << ellipse.AE();
   break; }

   case 'Q': break;
   default : cout << "Invalid choice. Try again " << endl;
   }
   } while(cmd != 'Q');
}  // end implementation of class Geometric


int main()
{
   Geometric g;    // g is an object of class Geometric
   g.menu();       // The dot operator . is used to access
  // a data member or member function of a
  // particular class object
   return 0;
}
```

4.7.2 Example 2.

```
// class1b.cpp
// calculates length of a point from the origin
// calculates distance of two points

#include <iostream.h>      // for cout
#include <math.h>          // for sqrt

     // Declaration of class Point
class Point
{
   private:
      double x, y;                        // private data members
   public:
      Point();                            // default constructor
      Point(const Point&);                // copy constructor
      ~Point();                           // destructor
      Point(double,double);               // constructor two arguments
      Point& operator = (const Point&);// assignment = overloaded
      double length(Point);               // member function
      double distance(Point,Point);       // member function
};

     // Implementation of class Point
Point::Point()
{
   cout << "default constructor called";
}

Point::Point(double x0,double y0)
{
   x = x0;  y = y0;
}

Point::Point(const Point& cp)
{
   x = cp.x;
   y = cp.y;
}

Point::~Point()
{
   cout << endl << "destructor called" << endl;
}
```

```
Point& Point::operator = (const Point& arg)
{
   x = arg.x;
   y = arg.y;
   return *this;
}

double Point::length(Point a)    // Point: abstract data type
{
   return sqrt(a.x*a.x + a.y*a.y);
}

double Point::distance(Point a,Point b)
{
   return sqrt((a.x-b.x)*(a.x-b.x)+(a.y-b.y)*(a.y-b.y));
}  // end implementation

int main()
{
   double norm;
   Point x(3.0,4.0);
   Point p;
   norm = p.length(x);
   cout << "norm = " << norm << endl;

   Point u(1.0,2.0);
   Point v(3.0,1.0);
   Point q;
   double dist = q.distance(u,v);
   cout << "distance = " << dist << endl;

   Point y(x);   // copy constructor called
   cout << "y.length(y) = " << y.length(y) << endl;

   Point z;
   z = x;   // overloaded assignment
   cout << "z.length(z) = " << z.length(z);

   return 0;
}
```

4.7.3 Example 3.

```
// nested.cpp
// nested classes

#include <iostream.h>   // for cout
#include <string.h>     // for strcpy

   // Declaration of class Records
class Records
{
   class Name
   {
      public:
        char first[40];
        char mid[40];
        char last[40];
   };
   Name N;                  // N is object of class Name

   class Phone
   {
      public:
        char country[4];
        char area[4];
        char number[8];
   };
   Phone P;                 // P is object of class Phone

   public:
      Records();   // default constructor

      Records(char*,char*,char*,char*,char*,char*);

      void display();
};

Records::Records() {  }

Records::Records(char* f,char* m,char* l,char* c,char* a,char* n)
{
   strcpy(N.first, f);
   strcpy(N.mid, m);
   strcpy(N.last, l);
   strcpy(P.country, c);
```

```
   strcpy(P.area, a);
   strcpy(P.number, n);
}

void Records::display()
{
   const char* SPACE = " ";
   cout << N.first << SPACE << N.mid << SPACE << N.last << endl;
   cout << "(" << P.country << ")";
   cout << SPACE << P.area << SPACE << P.number << endl;
}

int main()
{
   Records employee1("Jack","Lucky","Darwin","27","11","4892331");
   employee1.display();
   cout << endl;

   Records employee2("Sue","Jane","Cooper","1","231","3421099");
   employee2.display();

   Records employee3("Gene","Spirit","Smith","65","451","4512340");
   employee3.display();

   return 0;
}
```

4.7.4 Example 4.

The following example shows the overloading of the operators =, *=, += and []. The operator [] is overloaded for indexing.

```
// mvector.cpp

#include <iostream.h>
#include <assert.h>  // assert(i != 0); asserts a truth condition

class Vector
{
   private:
      int size;
      double* r;
   public:
      Vector();
      Vector(int);
      Vector(const Vector&);          // copy constructor
      ~Vector();
      double& operator [] (int);      // overloading []
      Vector& operator = (Vector&);
      Vector& operator *= (double&);
      Vector& operator += (Vector&);
};    // end declaration class Vector

Vector::Vector() {   }

Vector::Vector(int n)
{
   size = n;
   r = new double[size];
}

Vector::~Vector()
{
   delete [] r;
   cout << "Vector destructor called" << endl;
}

Vector::Vector(const Vector& v)
{
   size = v.size;
   assert(r != NULL);
   int i;
   for(i=0; i<size; i++)
```

```
   {
   r[i] = v.r[i];
   }
   cout << "copy constructor called" << endl;
}

double& Vector::operator [](int i)
{
   return r[i];
}

Vector& Vector::operator = (Vector& r2)
{
   size = r2.size;
   int i;
   for(i=0; i < size; i++)
   r[i] = r2[i];
   return *this;
}

Vector& Vector::operator *= (double& c)
{
   int i;
   for(i=0; i < size; i++)
   r[i] *= c;
   return *this;
}

Vector& Vector::operator += (Vector& r2)
{
   int i;
   for(i=0; i < size; i++)
   r[i] += r2[i];
   return *this;
}

int main()
{
   Vector v(2);
   v[0] = 1.0; v[1] = 3.0;

   int i;
   for(i=0; i<2; i++)
   {
   cout << "v[" << i << "] = " << v[i] << endl;
```

```
  }

  Vector w(2);
  w[0] = 3.1; w[1] = 4.7;

  w += v;      // w = w + v;
  for(i=0; i<2; i++)
  {
  cout << "w[" << i << "] = " << w[i] << endl;
  }

  Vector z(4);
  z[0] = v[1];  z[1] = w[0]; z[2] = 5.6;  z[3] = -1.3;

  Vector s(4);
  s = z;         // overloaded assignment operator called

  for(i=0; i<4; i++)
  {
  cout << "s[" << i << "] = " << s[i] << endl;
  }

  return 0;
}
```

4.7.5 Example 5.

In this example we overload the + operator for addition and the = for assignment. The function atof converts a string to a floating-point number. The function strtok searches one string for tokens, which are separated by delimiters defined in a second string. The function sprintf() writes formatted output to a string

```
// Symbol.cpp

#include <iostream.h> // for cout
#include <stdio.h>    // for sprintf
#include <string.h>   // for strtok
#include <math.h>     // for atof

   // Declaration of class Symbol
class Symbol
{
  private:
    double part1, part2;
  public:
    Symbol();                       // constructor no argument
    Symbol(const Symbol&);          // copy constructor
    Symbol(char*);                  // constructor one argument
    Symbol& operator = (const Symbol&); // overloading operator =
    Symbol operator + (const Symbol&);  // overloading operator +
    char* display();                // member function
};

   // Definition (Implementation) of class Symbol
Symbol::Symbol()
{
  part1 = part2 = (double) 0;
}

Symbol::Symbol(const Symbol& arg)
{
   part1 = arg.part1;
   part2 = arg.part2;
}

Symbol::Symbol(char* cnum)
{
   part1 = atof(strtok(cnum,"+"));
   part2 = atof(strtok(0,"b"));
}
```

```
Symbol& Symbol::operator = (const Symbol& arg)
{
   part1 = arg.part1;
   part2 = arg.part2;
   return *this;
}

Symbol Symbol::operator + (const Symbol& cnum)
{
  Symbol temp;
  temp.part1 = part1 + cnum.part1;
  temp.part2 = part2 + cnum.part2;
  return temp;
}

char* Symbol::display()
{
  char* temp = new char[10];
  sprintf(temp,"%fa + %fb",part1,part2);
  return temp;
}

int main()
{
  Symbol x("7a + -5b");
  Symbol y("8a + 9b");
  Symbol z;
  z = x + y;
  cout <<  "z = " << z.display() << endl; // z = 15.000a + 4.000b

  return 0;
}
```

4.8 Friend

A *friend F* of class X is a function or class that, although not a member function of X, has full access rights to the private and protected members of X. In all other respects, X is a normal function (or class) with respect to scope, declarations and definitions. Class friendship is not transitive: X friend of Y and Y friend of Z does not imply X friend of Z. However, friendship is inherited. Like a member function, a friend function is explicitly declared in the declaration of the class it is friend of and is thus as much part of the interface of that class as a member function is. The friend keyword is a function declaration modifier. It takes the general form

```
friend return_type function_name(parameters);
```

In the following program we give a simple example to explain friendship. In class X the entire class Y is proclaimed friend. Now all the member functions of class Y can access the private data of X. In the present program the single data item is pi. In the friend declaration we specify that Y is a class using the keyword friend class Y.

```
// mfriend.cpp

#include <iostream.h>

class X
{
   private:
     double pi;
   public:
     X();    // default construtor
     friend class Y;  // class Y is friend of class X
};

X::X()
{
   pi = 3.14159;
}

class Y
{
   private:
     double r;
   public:
     Y();
     Y(double);
```

```
   double f(X);
};

Y::Y() { }

Y::Y(double arg)
{
  r = arg;
}

double Y::f(X x)
{
  return r*r*x.pi;
}

int main()
{
  double a = 4.0;
  X z;
  Y v(a);
  double area = v.f(z);
  cout << "area = " << area;

  return 0;
}
```

4.9 Derivation and Inheritance

The class alone is not enough to support object-oriented programming and design. We would like to relate disparate classes to one another. We would use nested classes or friendship. Nested classes are, however, only a partial solution because nested classes do not have the required characteristics to fully relate one class to another. What we need is a mechanism for creating hierarchies of classes in which `class A` is a cognate of `class B`, but with added features. C++ has a mechanism for creating such a hierarchy the so-called derived class. The process of deriving classes is called inheritance. Inheritance is one of the distinguishing characteristics of object-oriented languages.

Object-oriented programming extends abstract data types to allow for type/subtype relationships. This is achieved through a mechanism referred to as inheritance. Rather than reimplementing shared characteristics, a class can inherit selected data members and member functions of other classes. In C++, inheritance is implemented through the mechanism of class derivation.

There is a defining mark in the definition of a derived class, which is found in the header of the definition:

```
class Derived1 : Base1
{
......
......
};
```

Recall that a `class` can be `private`, `protected` or `public`. If it is protected, its name can be used only by member functions and friends in which it is declared and by member functions and friends of classes derived from this `class`.
There are two important concepts in object-oriented programming, namely the concept of class hierarchy (an "Is A" relationship) and containership (a "Has A" relationship). The following example shows a hierarchy of two classes that illustrate successive refinement in a subclasses. It illustrates the "Is A" relationship using a Vehicle class and a Car class. Obviously a car is a vehicle.

```
// MVehicle.cpp

#include <iostream.h>

class Vehicle
{
   private:
    int weight;
    int topSpeed;
```

```
    float price;

   public:
    Vehicle();
    Vehicle(int,int,float);
    ~Vehicle();
    int getWeight();
    int getTopSpeed();
    float getPrice();
    void print();
};

Vehicle::Vehicle() {  }

Vehicle::Vehicle(int aWeight,int aTopSpeed,float aPrice)
{
   weight = aWeight;
   topSpeed = aTopSpeed;
   price = aPrice;
}

Vehicle::~Vehicle() {  }

int Vehicle::getWeight() {  return weight; }

int Vehicle::getTopSpeed() {  return topSpeed; }

float Vehicle::getPrice() {  return price; }

void Vehicle::print()
{
   cout << "\nWeight: " << weight << " kg";
   cout << "\nTop Speed: " << topSpeed << " km/h";
   cout << "\nPrice: " << price << " Dollar";
}

class Car : public Vehicle
{
   private:
    int numberCylinders;
    int horsepower;
    int displacement;

   public:
    Car();
```

```
    Car(int,int,float,int,int,int);
    ~Car();
    int getNumberCylinders();
    int getHorsepower();
    int getDisplacement();
    void print();
};

Car::Car() { }

Car::Car(int aWeight,int aTopSpeed,float aPrice,int aNumberCylinders,
         int aHorsepower,int aDisplacement)
         : Vehicle(aWeight,aTopSpeed,aPrice)
{
   numberCylinders = aNumberCylinders;
   horsepower = aHorsepower;
   displacement = aDisplacement;
}

Car::~Car()  { }

int Car::getNumberCylinders() { return numberCylinders; }

int Car::getHorsepower() {  return horsepower; }

int Car::getDisplacement() {  return displacement; }

void Car::print()
{
  Vehicle::print();
  cout << "\nCylinders: " << numberCylinders;
  cout << "\nHorsepower: " << horsepower;
  cout << "\nDisplacement: " << displacement << " cubic cm";
}

void main()
{
   Vehicle aVehicle(15000, 120, 30000.00);
   cout << "A vehicle: ";
   aVehicle.print();
   cout << endl << endl;
   Car aCar(3500,100,12000.00,6,120,300);
   cout << "A car: ";
   aCar.print();
}
```

4.10 Virtual Functions

Virtual functions allow a derived class to redefine member functions inherited from a base class. We can then write general programs that are oblivious to the specific classes of objects they manipulate through dynamic binding, the run time system will choose the function appropriate to a particular class. A base class can declare a member function using the keyword `virtual`, signifying that the base class's implementation of the member function is only a default which will be used only when a derived class does not supply its own implementation. If a derived class does supply its own implementation of a virtual member function, then that is the implementation that C++ will use when client programs apply the member function of instances of that class. Since a member function of the derived class implements the operation, it has no problem accessing the member variables and other member functions of the derived class. Moreover, when a client applies a virtual member function to an object, C++ automatically determines which class the object belongs to at execution time, and transfers control to the correct implementation of the function. This happens even when the client program treats the object as an instance of the base class. Ordinary member functions must be implemented by the class that declares them, so a base class would have to implement all of the operations that it provides for client programs. Member functions of a base class, however, cannot readily access the member variables and fuctions that might be added by derived classes, so a base class usually cannot implement member functions that are useful for instances of its derived classes. C++ gives us virtual member functions to solve these problems. A base class can declare a member function using the keyword virtual, signifying that the base class's implementation of the member function is only a default which will be used only when a derived class does not supply its own implementation.

Using C++ derived classes and virtual functions the program is more readable, general, and flexible than using `structs` with type fields and switch statements in C.

In the following example we calculate the area of a triangle, a rectangle and a polygon. Here we can use the virtual function `area()`. The area of a polygon with points

```
x[0], y[0],  x[1], y[1], .... , x[n-1], y[n-1]
```

is given by

$$A = \frac{1}{2}\sum_{i=0}^{n-1}(x[i] * y[i+1] - y[i] * x[i+1])$$

where $n \equiv 0$.

```
// virt.cpp

#include <iostream.h>
#include <math.h>

class Base
{
   public:
      virtual void area() = NULL;
};

class Derv1 : public Base
{
   public:
      void area();
};

void Derv1::area()
{
   double x1 = 2.0; double x2 = 3.2; double x3 = 2.8;
   double y1 = 1.5; double y2 = 3.5; double y3 = 0.5;
   double r = fabs((x1*y2-x2*y1+x2*y3-x3*y2+x3*y1-x1*y3)/2.0);
   cout << "area of triangle is: " << r << "\n";
}

class Derv2 : public Base
{
   public:
      void area();
};

void Derv2::area()
{
   double x1 = 0.5; double x2 = 1.5;
   double y1 = 0.4; double y2 = 1.9;
   double r = fabs((x1-x2)*(y1-y2));
   cout << "area of rectangle is: " << r << "\n";
}

class Derv3 : public Base
{
   public:
      void area();
};
```

```
void Derv3::area()
{
   double x[6], y[6];
   x[0] = 0.0;  y[0] = 0.0;  x[1] = 3.0;  y[1] = 0.0;
   x[2] = 2.0;  y[2] = 1.0;  x[3] = 3.0;  y[3] = 2.0;
   x[4] = 2.0;  y[4] = 2.0;  x[5] = 0.0;  y[5] = 1.0;
   double r = 0.0;
   for(int i=0; i<6; i++)
   {
   r += x[i]*y[((i+1)%6)] - y[i]*x[((i+1)%6)];
   }
   cout << "area of polygon is: " << r/2.0 << "\n";
}

int main()
{
   Derv1 dv1;   // triangle
   Derv2 dv2;   // rectangle
   Derv3 dv3;   // polygon
   Base *ptr;

   ptr = &dv1;          // address
   ptr -> area();       // output: area of triangle

   ptr = &dv2;          // address
   ptr -> area();       // output: area of rectangle

   ptr = &dv3;          // address
   ptr -> area();       // output: area of polygon

   return 0;
}
// The class object selector . is used with a class object
// or reference. The class pointer selector -> is used with
// a pointer to a class object.
```

Chapter 5

Function Templates

5.1 Introduction

By using function templates, we can define a pattern for a family of related overloaded functions by letting the data type itself be a parameter. Template functions provide a mechanism by which the programmer can take advantage of the compactness of the macro solution without forfeiting any of the benefits of a strongly-typed language. Traditionally, if one defines a *function* one would have to define it for every data type with which one wishes to be able to call that function. If one would write a sorting routine one would have to write the function for `char`, `short`, `int`, `long`, `float`, `double`, `string` etc.. If one wants to improve the algorithm at a later stage one would have to search for all definitions of the function and make the relevant changes. This is cumbersome and prone to errors. By defining a function on a template the compiler will automatically generate the code for the data types required in the code.

The `template` keyword always begins both a definition and a forward declaration of a template function. This is followed by a comma-separated list of formal type parameters bracketed by the less-than `<` and greater-than `>` tokens. This list is spoken of as the formal parameter list of the template. It cannot be empty. Each formal parameter consists of the keyword `class` followed by an identifier. The keyword `class` indicates that the parameter represents a potential built-in or user-defined type. The name of a formal parameter can occur only once within the template parameter list. The function definition or declaration follows the formal template parameter list. An example is

```
template <class Type> Type min (Type x, Type y)
```

or

```
template <class T> T min (T x, T y)
```

In the following programs we give some examples of function templates.

5.2 Examples

A Bubble sort algorithm defined on a template. It can hence sort a vector of any class, integers, floating point numbers, characters, or any user-defined class (e.g. a data base of employees)

```
// temp1.cpp
#include <iostream.h>

template <class T>                        // T is determined by
void BubbleSort (T vec[],int length)      // the calling program
{
  int exchanged, nleft = length;

  do
  {
    exchanged = 0;
    nleft--;
    int n;
    for(n=0; n<nleft; n++)
    {
      if (vec[n] > vec[n+1])  // if next element smaller: exchange
      {
      T dummy   = vec[n+1];
      vec[n+1] = vec[n];
      vec[n]   = dummy;
      exchanged++;
      }
    }
  } while(exchanged);  // if no exchanges in previous loop: done
}

int main()
{
  int lngth=5;
  float* fvec = new float[lngth];    // create a vector of floats

  fvec[0] = 8.1;
  fvec[1] = 3.7;
  fvec[2] = 4.2;
  fvec[3] = 9.0;
  fvec[4] = 7.2;

  BubbleSort(fvec,lngth);      // bubble sort this vector
```

```
  int i;
  for(i=0; i<lngth; i++)     // show result on screen
  cout << "fvec[" << i << "] = " << fvec[i] << endl;

  delete[] fvec;   // vector no longer needed => discard

  cout << endl;

  lngth=6;                          // the same for vector of chars
  char* cvec = new char[lngth];

  cvec[0] = 's';
  cvec[1] = 'd';
  cvec[2] = 'r';
  cvec[3] = 'a';
  cvec[4] = 'z';
  cvec[5] = 'm';

  BubbleSort(cvec,lngth);

  for(i=0; i<lngth; i++)
  cout << "cvec[" << i << "] = " << cvec[i] << endl;

  delete[] cvec;

  return 0;
}
```

In the following example we find the position of a given element in an array. If the element is not in the array we return −1.

```
// temp2.cpp

#include <iostream.h>

template <class T> int find(T* array,int n,const T& x)
{
   T* p = array;
   int i;
   for(i=0; i < n; i++)
   {
   if(*p == x)
   return i;
   p++;
   }
   return -1;
}

int main()
{
   int res1;
   int n = 4;
   int* a = NULL;
   a = new int[n];
   a[0] = 4; a[1] = 7; a[2] = 11; a[3] = 3;

   res1 = find(a,n,11);
   cout << "res1 = " << res1 << endl; // => 2
   delete [] a;

   int res2;
   n = 5;
   double* b = NULL;
   b = new double[n];
   b[0] = 1.7; b[1] = 3.1; b[2] = 3.7; b[3] = 0.4; b[4] = 7.1;

   res2 = find(b,n,0.4);
   cout << "res2 = " << res2 << endl; // => 3
   delete [] b;

   return 0;
}
```

```
// temp3.cpp

#include <iostream.h>   // for cout

template <class T> T Min(T *a,const int n)
{
   T min_val = a[0];
   int j;
   for(j=1;j<n;++j)
   if(a[j] < min_val)
   min_val = a[j];

   return min_val;
}

int main()
{
   const int n = 4;
   int x[n];
   x[0] = 2; x[1] = 7; x[2] = -3; x[3] = 5;
   int res1 = Min(x,n);
   cout << res1 << "\n";  // => -3

   const int m = 5;
   float y[m];
   y[0] = 1.3; y[1] = 2.7; y[2] = 0.7; y[3] = 1.7; y[4] = 7.2;
   float res2 = Min(y,m);
   cout << res2 << "\n";  // => 0.7

   const int r = 3;
   char z[r];
   z[0] = 'c'; z[1] = 'f'; z[2] = 'b';
   char res3 = Min(z,r);
   cout << res3 << "\n"; // => b

   return 0;
}
```

```
// temp4a.cpp
// two generic types

#include <iostream.h>

template <class T1, class T2> T2 plus(T1 x,T2 y)
{
   return (x + y);
}

int main()
{
   int a = 2;
   double b = 3.14;
   cout << plus(a,b) << endl;  // => 5.14
   return 0;
}
```

```
// temp4b.cpp

#include <iostream.h>

template <class T1, class T2> T1 plus(T1 x,T2 y)
{
   return (x + y);
}

int main()
{
   int a = 2;
   double b = 3.14;
   cout << plus(a,b) << endl;  // => 5
   return 0;
}
```

```
// temp5.cpp

#include <iostream.h>

template <class T> void printArray(T *array,const int count)
{
   int i;
   for(i = 0; i < count; i++)
   cout << array[i] << " ";
   cout << endl;
}

int main()
{
   const int aCount = 5;
   const int bCount = 7;
   const int cCount = 6;
   int A[aCount] = { 2, 3, 5, 7, 11 };
   float B[bCount] = { 0.3, 0.5, 0.9, 1.2, 1.7, 2.1, 3.3 };
   char C[cCount] = "Hello"; // 6-th position for null character

   cout << "array A contains: " << endl;
   printArray(A, aCount);

   cout << "array B contains: " << endl;
   printArray(B, bCount);

   cout << "array C contains: " << endl;
   printArray(C, cCount);

   int length;
   cout << "enter length of string: ";
   cin >> length;
   char *str = NULL;
   str = new char [length+1];
   cout << "enter string: ";
   cin >> str;
   cout << "string contains: ";
   printArray(str,length);

   delete str;

   return 0;
}
```

The `sizeof` operator returns the size, in bytes, of an expression or type specifier. Application of the `sizeof` operator on a pointer type returns the size of the memory necessary to contain an address of that type.

```
// temp6.cpp

#include <iostream.h>

template <class T1, class T2, class T3>
int Size(T1 a, T2 b, T3 c)
{
   cout << "a = " << a << endl;     // output 49
   cout << "b = " << b << endl;     // output 3.14
   cout << "c = " << c << endl;     // address
   int x = sizeof(a);  cout << "x = " << x << endl; // 4
   int y = sizeof(b);  cout << "y = " << y << endl; // 8
   int z = sizeof(c);  cout << "z = " << z << endl; // 4
   return (x > y && x > z ? x:(y > z ? y:z));
}

int main()
{
  cout << endl << endl;
  int i = 49;  int* pi;  pi = &i;
  cout << "size of i: " << sizeof(i) << endl;   // 4
  cout << "size of pi: " << sizeof(pi) << endl; // 4

  double d = 3.14;  double* pd;  pd = &d;
  cout << "size of d: " << sizeof(d) << endl;   // 8
  cout << "size of pd: " << sizeof(pd) << endl; // 4

  int str[7] = { 17, 19, 23, 29, 31, 37, 41 };
  cout << "size of str: " << sizeof(str) << endl;    // 28

  int res = Size(i,d,str);
  cout << "res = " << res << endl;  // 8

  return 0;
}
```

Chapter 6

Class Templates

6.1 Introduction

A class template (also called a generic class or class generator) lets us define a pattern for class definitions. The forward declaration of the template `Myclass` class is as follows

```
template <class T> class Myclass
```

The `template` keyword begins every forward declaration and definition of a template class. It is followed by a formal template list surrounded by angle brackets (< >). The formal parameter list cannot be empty. Multiple parameters are separated by commas

```
template <class T1, class T2, class T3, class T4>
```

Declarations such as

```
template <class T, int size> class Myclass
```

are also allowed. When the C++ compiler encounters the object definitions, it first substitutes the type names for the `T` symbol within the class. If one class template uses a second template, the compiler will perform type substitutions one class at a time, beginning with the first class referenced.

We consider four examples. Our first example is a vector class (a one-dimensional array). Whether we have a vector of integers, complex numbers, characters, strings or any other type (for example rational numbers, quaternions, etc.) the basic operations performed are the same (insert, delete, index, compare and so on). With the element type treated as a type parameter to the class, the system will generate type-safe class definitions. Our second example is a stack class. In our third example we also consider a stack class template which includes nodes to store and link stack elements. In the fourth example we consider the case with two generic data types.

6.2 Examples

6.2.1 A Template Vector Class

```
// vektor.cpp
// Application of class templates

#include <iostream.h>
#include <string.h>


      // definition of class Vector
template <class T> class Vector
{
  T *data;
  int size;

  public:
    Vector();
    Vector(int);
    Vector(const Vector&);
    ~Vector();
    Vector<T>& operator = (const Vector<T>&);
    T& operator[](int);
    int length();
    void resize(int);
};

// implementation of class Vector
template <class T> Vector<T>::Vector() {   }

template <class T> Vector<T>::Vector(int n)
{
   data = new T[n];
   size = n;
}

template <class T> Vector<T>::Vector(const Vector& v)
{
   int i;
   size = v.size;
   for(i=0; i<size; i++)
   {
   data[i] = v.data[i];
   }
```

```
}

template <class T> Vector<T>::~Vector()
{
  delete [] data;
}

template <class T>
Vector<T>& Vector<T>::operator = (const Vector<T>& v)
{
   size = v.size;
   if(this != &v)  // check for self assignment
   {
   resize(v.length());
   int i;
   for(i=0; i<size; i++)
   data[i] = v[i];
   }
   return *this;
}

template <class T> T& Vector<T>::operator[] (int i)
{
   return data[i];
}

template <class T> int Vector<T>::length()
{
   return size;
}

template <class T> void Vector<T>::resize(int length)
{
   delete [] data;
   data = new T[length];
}

void main()
{
   Vector<int> x(5);  // generates a vector of integers
   // The symbol Vector must always be accompanied by a
   // data type in angle brackets
   int i;
   for(i=0;i<5;++i)
   x[i] = i*i;
```

```
for(i=0;i<5;++i)
cout << x[i] << "\n";

Vector<double> u(100); // generates a vector of double
u[0] = 0.618;
for(i=1;i<10;i++)
{
u[i] = 4.0*u[i-1]*(1.0-u[i-1]);
}

Vector<char> ch(3);  // generates a vector of characters
ch[0] = 'a'; ch[1] = 'b'; ch[2] = 'c';
cout << "\n" << ch[1] << "\n";
cout << ch[1]-ch[2] << "\n";

Vector<char *> c(2); // generates a vector of strings
c[0] = "aba"; c[1] = "bab";
for(i=0;i<2;++i)
cout << c[i] << "\n";
strcpy(c[0],c[1]);
cout << c[0] << "\n";

Vector<char *> C(5);
char *str;
str = "moses";

C[0] = "aaron"; C[1] = "berta"; C[2] = "claus";
C[3] = "catherine"; C[4] = "moses";
int result[5];
int p = -1;
for(i=0;i<5;i++)
{
result[i] = strcmp(C[i],str);
}
for(i=0;i<5;i++)
{
cout << result[i] << "\n";
if(result[i] == 0) p = i;
}
cout << "\n" << " p = " << p;
if(p != -1) cout << " name is in list " << "\n";
else
cout << " name is not in list " << "\n";

Vector<double> X(100), Y(100); // generates two vectors of double
```

```
   Vector<char> PP(100); // vector of characters for symbolic dynanmics
   X[0] = 0.1;  Y[0] = 0.3;
   for(i=1;i<20;i++)
   {
   X[i] = 1.0 - 1.4*X[i-1]*X[i-1] + Y[i-1];
   Y[i] = 0.3*X[i-1];
   if(X[i] > 0 && Y[i] > 0)
   PP[i] = 'p'; else PP[i] = 'x';
   }
   cout << "PP[6] = " << PP[6] << "\n";
   cout << "PP[17] = " << PP[11] << "\n";

   Vector<char[15]> M(2);
   strcpy(M[0],"object-");
   strcat(M[0],"oriented");
   cout << "M[0] = " << M[0] << endl;
   strcpy(M[1],"abba");
   int b = strcmp(M[0],M[1]);
   cout << "b = " << b << endl;

   Vector<Vector<int> > vec(3);

   for(i=0; i<=2; i++)
   vec[i].resize(4);

   vec[0][0] = 7;  vec[0][1] = 4;  vec[0][2] = -3;  vec[0][3] = -17;
   vec[1][0] = 4;  vec[1][1] = 9;  vec[1][2] = 0;   vec[1][3] = 5;
   vec[2][0] = 7;  vec[2][1] = 11; vec[2][2] = 8;   vec[2][3] = 77;
   for(i=0; i<=2; i++)
   {
   int j;
   for(j=0; j<=3; j++)
   cout << vec[i][j] << "  ";
   cout << endl;
   }
   cout << "vec[0][0] = " << vec[0][0] << endl;
   int res = vec[0][0]*vec[2][3];
   cout << "res = " << res << endl;

   for(i=0; i<=2; i++)
   vec[i].resize(0);
}
```

6.2.2 A Template Stack Class

```
// stackt.cpp

#include <iostream.h>
#include <iomanip.h>    // for setiosflags

     // Declaration of class Stack
template <class T> class Stack
{
   private:
      T *storage;
      int elements;
      int isempty;
      int isfull;
      int stack_size;
   public:
      Stack(int);      // constructor
      T pop();
      T push(T value);
      int is_empty();
      int is_full();
};

     // Definition of class Stack
template <class T> Stack<T>::Stack(int size)
{
   storage = new T[size];
   elements = 0;
   isempty = 1;
   isfull = 0;
   stack_size = size;
}

template <class T> T Stack<T>::pop()
{
   if(is_empty())
   return 0;
   else
   {
   if(--elements == 0)
   isempty = 1;
   isfull = 0;
   return storage[elements];
   }
```

```
}

template <class T> T Stack<T>::push(T value)
{
   if(is_full())
   return 0;
   else
   {
   isempty = 0;
   storage[elements++] = value;
   if(elements == stack_size)
   isfull = 1;
   return value;
   }
}

template <class T> int Stack<T>::is_empty()
{
   return isempty;
}

template <class T> int Stack<T>::is_full()
{
   return isfull;
}

int main()
{
   Stack<int> values(10);
   Stack<float> prices(15);

   int i;
   for(i = 0; ! values.is_full(); i++)
   values.push(i);
   while(! values.is_empty())
   cout << values.pop() << endl;
   for(i = 0; ! prices.is_full(); i++)
   prices.push(i * 100.0);
   while(! prices.is_empty())
   cout << setiosflags(ios::showpoint) << prices.pop() << endl;
   return 0;
}
```

6.2.3 A Template Stack Class with Node

```
// stacktem.cpp

#include <iostream.h>

template <class T> class Stack
{
   struct Link
   {
      T element;
      Link *next;
      Link()
      { next = NULL; }
      Link(T elem, Link *nptr = NULL)
      {
      element = elem;
      next = nptr;
      }
      ~Link()
      {  delete next; }
   };

   protected:
      Link* top;
   public:
      Stack();
      ~Stack();
      void push(T elem);
      T pop();
      int empty() const;
      int full() const;
      friend ostream& operator << (ostream& out, const Stack<T>& s);
};

 // Definition of class Stack
template <class T> Stack<T>::Stack()
{
   top = NULL;
}

template <class T> Stack<T>::~Stack()
{
   delete top;
   top = NULL;
```

```
}

template <class T> void Stack<T>::push(T elem)
{
   top = new Link(elem,top);
}

template <class T> T Stack<T>::pop()
{
   T elem = top -> element;
   Stack<T>::Link* nodeptr;
   nodeptr = top;
   top = top -> next;
   nodeptr -> next = NULL;
   delete nodeptr;

   return elem;
}

template <class T> int Stack<T>::empty() const
{
   return top == NULL;
}

template <class T> int Stack<T>::full() const
{
   return 0;
}

template <class T>
ostream& operator << (ostream& out,const Stack<T>& st)
{
   Stack<T>::Link*  nodeptr;
   nodeptr = st.top;

   while(nodeptr != NULL)
   {
   out << nodeptr -> element << endl;
   nodeptr = nodeptr -> next;
   }

   return out;
}

ostream& displaybinary(unsigned num,ostream& out = cout);
```

```
int main()
{
   cout << endl << endl;

   Stack<int> intval;
   Stack<float> floatval;
   intval.push(7);
   floatval.push(11.1);
   intval.push(13);
   floatval.push(17.3);
   intval.push(19);
   cout << intval.pop() << endl;     // output 19
   cout << floatval.pop() << endl;   // output 17.299999
   cout << intval.pop() << endl;     // output 13
   cout << floatval.pop() << endl;   // output 11.1
   cout << intval.pop() << endl;     // output 7
   cout << endl;

   cout << "enter a string: ";
   char ch;
   Stack<char> charstack;

   for(;;)
   {
   cin.get(ch);
   if(ch == '\n')
   break;
   charstack.push(ch);
   }
   cout << charstack << endl << endl;

   cout << "enter an integer: ";
   unsigned number;
   cin >> number;
   displaybinary(number);
   return 0;
}
```

```
ostream& displaybinary(unsigned num,ostream& out)
{
   Stack<int> intstack;

   while(num != 0)
   {
   intstack.push(num % 2);
   num /= 2;
   }
   while(!intstack.empty())
   out << intstack.pop();

   return out;
}
```

The output is

```
19
17.299999
13
11.1
7
enter a string: willi
i
l
l
i
w
enter an integer: 333
101001101
```

6.2.4 Two Generic Data Types

The following example shows how to use more than one generic data type.

```
// twogen.cpp

#include <iostream>
#include <string>
using namespace std;

template <class T1, class T2> class TwoGeneric
{
  private:
    T1 x;
    T2 y;
  public:
    TwoGeneric();
    TwoGeneric(T1, T2);
    void display();
};

template <class T1,class T2> TwoGeneric<T1,T2>::TwoGeneric()
{ }

template <class T1,class T2> TwoGeneric<T1,T2>::TwoGeneric(T1 u,T2 v)
{
   x = u; y = v;
}

template <class T1,class T2> void TwoGeneric<T1,T2>::display()
{
   cout << x << "  " << y << endl;
}

int main()
{
  TwoGeneric<int,double> ob1(-30, 3.14159);
  TwoGeneric<char,string> ob2('+', "Egoli");

  ob1.display();
  ob2.display();

  return 0;
}
```

Chapter 7

Sorting and Searching

7.1 Introduction

Sorting is a process by which a collection of items is placed into ascending or descending order on one of the fields in the items. This field is called a *key* . Sometimes the items are structures, in the usual C sense, with one of the fields being the key. Other times, especially for studying sorting algorithms, each item is just an integer, with the collection being an array. In this case we can say that the task of sorting is to produce a permutation of the array elements that places them in ascending order. Allowing for the possibility of duplicates, we can make the statement broader by saying non-descending order instead of ascending order. All the standard sorting methods are introduced such as bubblesort, quicksort, and shellsort. Both number sorting and name sorting are performed. Using class and function templates we can unify number and name sorting. Furthermore, we sort vectors according to their length.

Searching is a process in which we look for, locate, and retrieve one item from a table of items. It is one of the most common uses of a computer. For example, we search to see whether a word is in a dicionary. The collection in which we search is called either a file of records or a table of items. Each entry in a table has at least one key field by which it is identified. It also usually has at least one non-key field, and may have many. We also introduce the divide-and-conquer methodology.

Sorting and searching algorithms are described in detail by Baase [2].

7.2 Largest Number in an Array

The following program finds the largest (integer) number of an array typed in. The maximum size of the array is 20.

```
// maxnum.cpp

#include <iostream.h>   // for cout
#include <stdio.h>      // for scanf()

int Max(int[], int);    // prototype

void main()
{
   const int MAXSIZE = 20;  // maximum size of array
   int list[MAXSIZE];
   int size = 0;           // start at element [0]
   int num;                // temporary storage
   do                      // get list of numbers
   {
   cout << " enter integer numbers: ";
   scanf("%d", &list[size]);
   }
   while(list[size++] != 0);  // exit loop on 0
   num = Max(list,size-1);    // get largest number
   cout << "\n" << " largest number is: " << num << "\n";
}

int Max(int list[], int size)
{
   int t, max;
   max = list[0];               // assume first element is largest
   for(t=1; t<size; t++)        // checks remaining elements
   if(max < list[t])            // if one bigger,
   max = list[t];               // make it the largest
   return (max);
}
```

7.3 Quicksort and Bubblesort

The following program sorts numbers by applying quicksort and bubblesort using the function swap. Bubblesort sorts by making several passes through the list, comparing pairs of keys in adjacent locations and interchanging them if they are out of order. That is, the first and second keys are compared and interchanged if the first is larger than the second; then the (new) second and the third key are compared and interchanged if necessary, and so on. Bubblesort is an inefficient N^2 algorithm, where N is the number of elements to be sorted. Quicksort is a partition-exchange sorting method: by pairwise exchanges of elements, the original array is partitioned into two subarrays. At the end of a round of partitioning, the partitioning element a is in its final place in the array. All elements in the left subarray are $\leq$ a, while all elements in the right subarray are $\geq$ a. The process is then repeated on the left and right subarrays independently, and so on until every partition consists of a single item only.

```
// quick.cpp

#include <iostream.h>

static void swap(int *ia,int i,int j) // swap two integers
{
     int t = ia[i];
     ia[i] = ia[j];
     ia[j] = t;
}

void bsort(int* ia,int sz )     // bubblesort
{
     int ix, j;
     for(ix = 0;ix< sz;ix++ )
     for(j=ix+1;j<sz;j++)
     if(ia[ix] > ia[j])
     swap(ia,ix,j);
}

void qsort(int *ia,int low,int high)  // quicksort
{
   if(low < high)
   {
   int lo = low;
   int hi = high + 1;
   int elem = ia[low];
```

```
  for(;;)
  {
  while(ia[++lo] < elem);
  while(ia[--hi] > elem);

  if(lo < hi)
  swap(ia,lo,hi);
  else
  break;
  }  // end for(;;)

  swap(ia,low,hi);
  qsort(ia,low,hi-1);      // recursion
  qsort(ia,hi+1,high);
  } // end if(low < high)
}

// two integer arrays to be sorted
int ia1[10] = { 26, 5, 37, 1, 61, 11, 59, 15, 48, 17 };

int ia2[12] =
{ 503, 87, 512, 61, 908, 170, 897, 275, 653, 426, 54, 9 };

int main()
{
  cout << "Bubblesort of first array " << endl;
  bsort(ia1,10);
  int k;
  for(k=0;k<10;k++)
  cout << ia1[k] << "\n";

  cout << "Quicksort of second array " << endl;
  qsort(ia2,0,11);
  int n;
  for(n=0;n<12;n++)
  cout << ia2[n] << "\n";

  return 0;
}
```

7.4 Shellsort

The program `shellsort` sorts an array of integers using the *shellsort* method. The basic idea of this sorting algorithm is that in early stages, far-apart elements are compared, rather than adjacent ones as in simpler interchange sorts. This tends to eliminate large amounts of disorder quickly, so that later stages have less work to do. The interval between compared elements is gradually decreased to one, at which point the sort effectively becomes an adjacent interchange method.

In the program `shell1.cpp` we have a list of students together with their age and sex. We sort this list with respect to their age. The sorting algorithm is shellsort and it is implemented as a function template. The operations on the employees is implemented in the `class Students`. The assignment operator = is overloaded.

```
// shellsort.cpp

#include <iostream.h>

void ss(int v[],int n)
{
   int d, i, j, t;
   for(d=n/2;d>0;d /=2)
   for(i=d;i<n;i++)
   for(j=i-d; j>=0 && v[j]>v[j+d]; j-=d)
   {
   t = v[j];
   v[j] = v[j+d];
   v[j+d] = t;
   }
}

void main()
{
   const int n = 5;
   int v[n];
   v[0] = 2; v[1] = 7; v[2] = -13; v[3] = 4; v[4] = 9;
   ss(v,n);

   int k;
   for(k=0;k<n;k++)
   cout << "\n" << "v[" << k << "] = " << v[k];
}
```

```
// shell1.cpp

#include <iostream.h>
#include <string.h>

// function template shell
template <class T> void shell(T* a,int st,int fn)
{
   unsigned long i, j, h;
   T item;
   h = 1;
   do
   {
   h = 3*h + 1;
   } while(h < fn-st+1);

   do
   {
   h /= 3;
   for(i=st+h; i<=fn; i++)
   {
   item = a[i];
   j = i;
   while(item < a[j-h])
   {
   a[j] = a[j-h];
   j -= h;
   if(j < st+h)
   break;
   }
   a[j] = item;
   }
   }  while(h > 1);
}

class Students
{
   protected:
      int age;
      char c;          // for sex: F for female M for male
      char name[16];
   public:
      Students();
      Students(char*,int,char);
   Students(const Students&);
```

```
      Students& operator = (Students&);
      friend int operator < (Students&,Students&);
      friend ostream& operator << (ostream&,Students&);
};

Students::Students()
{
   name[0] = '\0';
   age = 0;
   c = '\0';
}

Students::Students(char* n,int a,char ch)
{
   strcpy(name,n);
   age = a;
   c = ch;
}

Students(const Students& p)
{
   strcpy(name,p.name);
   age = p.age;
   c = p.c;
}

Students& Students::operator = (Students& p)
{
   strcpy(name,p.name);
   age = p.age;
   c = p.c;
   return *this;
}

int operator < (Students& x,Students& y)
{
   return x.age < y.age;
}

ostream& operator << (ostream& s,Students& p)
{
   s << p.name << " " << p.c << " " << p.age;
   return s;
}
```

```
int a1[5] = { 13, 82, 34, 93, 11 };

Students a2[6] =
{
   Students("Hugo", 22, 'M'),
   Students("Otto", 28, 'M'),
   Students("Jane", 43, 'F'),
   Students("Marc", 45, 'M'),
   Students("Sue", 33, 'F'),
   Students("Helen", 39, 'F')
};

int main()
{
   shell(a1,0,4);
   int i;
   for(i = 0; i < 5; i++)
   cout <<  "a[" << i << "] = " << a1[i] << endl;

   shell(a2,0,5);
   for(i=0; i < 6; i++)
   cout << "a2[" << i << "] = " << a2[i] << endl;

   return 0;
}
```

7.5 Using the Function bsearch

The function bsearch searches a table (array) of n elements in memory, and returns the address of the first entry in the table that matches the search key. The array must be in order.

```
// search1.cpp

#include <iostream.h>
#include <stdlib.h>

typedef int(*fptr) (const void*,const void*);

#define NELEMS(arr) (sizeof(arr)/sizeof(arr[0]))

int numarray[] = { 123, 145, 512, 627, 800, 933 };

int numeric(const int *p1,const int *p2)
{  return(*p1 - *p2); }

#pragma argsused
int lookup(int key)
{
   int *itemptr;
   // The cast of (int(*)(const void*,const void*)) is needed
   // to avoid a type mismatch error at compile time
   itemptr = (int*) bsearch (&key,numarray,NELEMS(numarray),
      sizeof(int),(fptr)numeric);
   return (itemptr != NULL);
}

void main()
{
   int i;
   cout << " enter integer number: ";
   cin >> i;
   if(lookup(i))
     cout << i << " is in the table" << "\n";
   else
     cout << i << " isn't in the table" << "\n";
}
```

7.6 Using the Function lsearch

The function `lsearch` searches a table for information. Since this is a linear search, the table entries do not need to be sorted before `lsearch` is called. If the item that key points to is not in the table, `lsearch` appends that item to the table.

```
// search2.cpp

#include <iostream.h>
#include <stdlib.h>
#include <string.h>     // for strcmp()

char* colors[10] = { "red", "blue", "green" };
int ncolors = 3;

int colorscmp(char **arg1,char **arg2)
{
   return (strcmp(*arg1,*arg2));
}

int addelem(char *key)
{
   int oldn = ncolors;
   lsearch(key, colors,(size_t *) &ncolors,sizeof(char *),(int(*)
      (const void *,const void *)) colorscmp);
   return(ncolors == oldn);
}

void main()
{
   int i;
   char *key = "purple";
   if(addelem(key))
     cout << " already in color table: " << key << "\n";
   else {
     strcpy(colors[ncolors-1],key);
     cout << " added to color table: " << key << "\n";
   }
   cout << " The colors are: " << "\n";
   for(i=0;i < ncolors; i++)
     cout << colors[i] << "\n";
}
```

7.7 Name Sorting

Here we give two programs that sort names (strings). The programs take a series of names typed in by the user, places them in an array, and sorts the pointers to the array so that, in effect, the names are rearranged into alphabetical order. As output the sorted list is given.

When we pass an array to a function, we are only passing its address. The array itself does not move. The same is true of strings: when we pass a string to a function, we are passing only its address. This ability to reference strings by addresses can be very helpful when we want to manipulate a group of strings. We sort an array of pointers to the strings. The program `sort1.cpp` uses both an array of strings and an array of pointers. The pointers occupy the array declared by the expression

```
char *ptr[MAXNUM];
```

The strings occupy a two-dimensional array

```
static char name[MAXNUM][MAXKEN];
```

The functions `strlen` and `strcmp` are used. The functions `strlen` determines the length of a string. The function `strcmp` performs an unsigned comparision of the two strings starting with the first character in each string and continuing with subsequent characters until the corresponding characters differ or until the end of the strings is reached.

```
// sort1.cpp

#include <iostream.h>
#include <stdio.h>      // for gets
#include <string.h>     // for strlen, strcmp

void main()
{
   const int MAXNUM = 30;
   const int MAXLEN = 51;

   static char name[MAXNUM][MAXLEN];
   char *ptr[MAXNUM];
   char *temp;
   int count = 0;
   int in, out;

   while(count < MAXNUM)
   {
   cout << " Name " << (count+1) << ": ";
   gets(name[count]);
   if(strlen(name[count])==0) break;
   ptr[count] = name[count];
   count++;
   }

   for(out=0;out<count-1;out++)
   for(in=out+1;in<count;in++)
   if(strcmp(ptr[out],ptr[in]) > 0 )
   {
   temp = ptr[in];
   ptr[in] = ptr[out];
   ptr[out] = temp;
   }

   cout << "\n" << " Sorted list: " << "\n";
   for(out=0;out<count;out++)
   cout << " Name " << out+1 << " " << ptr[out] << "\n";
}
```

```
// sort2.cpp

#include <iostream.h>
#include <string.h>          // for strcmp

class Names
{
   protected:
      char name[40];
   public:
      void setName();
      void printName();
      char* getName();
};

void Names::setName()
{
   cout << "Enter name: ";
   cin >> name;
}

void Names::printName()
{
   cout << endl << name;
}

char* Names::getName()
{
   return name;
}

void Sort(Names**,int);
void order(Names**,Names**);

int main()
{
   Names* persPtr[100];
   int n = 0;
   char choice;

   do
   {
   persPtr[n] = new Names;
   persPtr[n] -> setName();
   n++;
```

```
   cout << "Insert another (y/n)? ";
   cin >> choice;
   }
   while(choice=='y');
   cout << "\nUnsorted List;";
   int j;
   for(j=0; j<n; j++)
   {
   persPtr[j] -> printName();
   }

   Sort(persPtr,n);

   cout << "\nSorted List:";
   for(j=0; j<n; j++)
   {
   persPtr[j]->printName();
   }

   return 0;
}    // end main program

void Sort(Names** pp,int n)
{
   int j, k;
   for(j=0; j<n-1; j++)
   for(k=j+1; k<n; k++)
   order(pp+j,pp+k);
}

void order(Names** pp1,Names** pp2)
{
   if(strcmp((*pp1)->getName(),(*pp2)->getName()) > 0)
   {
   Names* tempptr = *pp1;
   *pp1 = *pp2;
   *pp2 = tempptr;
   }
}
```

7.8 Name and Number Sorting Using Templates

Using templates we can write a program that sorts numbers and names together, where the sorting algorithm need only be implemented once. The program contains the class

```
class Person
```

where the constructor

```
Person(char*)
```

takes the name of the person. We overload the operators

```
  <    ==    >
```

so that we can compare strings.

The program `sort3.cpp` includes the two template functions `sort` and `find`. In `sort` we sort the names and numbers using the operators `<` and `>`. Using the template function `find` we find out whether the name is in the data base provided by the `class Vector`. The `class Person` is used as an abstract data type in `class Vector`, i.e.

```
Vector<Person> V(3);
```

Thus the template function `sort` can be used to sort numbers and names, since the operator `>` and `<` are overloaded. The template function `find` finds names and numbers in a vector.

The function `sort` uses the inefficient bubblesort, which should be replaced by quicksort, shellsort or mergesort.

The program `sort3.cpp` also contains a data base of three names

```
Peter   Rudi   Ralf
```

The program asks for a name. It then checks whether the name is in the data base. Moreover, the names in the data base are sorted. Both classes `Person` and `Vector` contain a destructor.

```
// sort3.cpp

#include <iostream.h>
#include <string.h>   // for strlen, strcpy

    // Declaration of class Person
class Person
{
   protected:
     char *name;
   public:
     Person();
     Person(char*);
     Person(const Person&);
     ~Person();
     Person& operator = (const Person&);
     friend int operator < (Person&,Person&);
     friend int operator > (Person&,Person&);
     friend int operator == (Person&,Person&);
     friend ostream& operator << (ostream&,Person&);
};

Person::Person() { name = NULL; }

Person::Person(char *c)
{
   name = new char[strlen(c)+1];
   strcpy(name,c);
}

Person::Person(const Person& P)
{
   name = new char[strlen(P.name) + 1];
   strcpy(name,P.name);
}

Person::~Person() { delete [] name; }

Person& Person::operator = (const Person& P)
{
   if(this != &P)
   {
   delete [] name;
   name = new char[strlen(P.name) + 1];
   strcpy(name,P.name);
```

```
   }
   return *this;
}

int operator < (Person& a,Person& b)
{ return strcmp(a.name,b.name) < 0; }
int operator > (Person& a,Person& b)
{ return strcmp(a.name,b.name) > 0; }
int operator == (Person& a,Person& b)
{ return strcmp(a.name,b.name) == 0; }
ostream& operator << (ostream& s,Person& P) { s << P.name; return s; }

template<class T> class Vector
{
   protected:
      T *vec;
      int dim;
   public:
      Vector(int);
      ~Vector();
      int get_dim();
      T& operator [] (int);
      void print();
};

template<class T> Vector<T>::Vector(int n)
{
   dim = n;  vec = new T[n];
}

template<class T> Vector<T>::~Vector() { delete [] vec; }

template<class T> int Vector<T>::get_dim() { return dim; }

template<class T> T& Vector<T>::operator [] (int pos) { return vec[pos]; }

template<class T> void Vector<T>::print()
{
   int j;
   for(j=0;j<dim;j++) cout << vec[j] << endl;
}

// function template
template<class T> void find(Vector<T>& a,T& b)
{
```

```
   int j;
   for(j=0;j<a.get_dim();j++)
   {
   if(b==a[j])
   {
   cout << a[j] << " found in data base " << endl;
   break;
   }
   if(j==a.get_dim() - 1)
   cout << " sorry! Not found in data base " << endl;
   }
}

// function template
template<class T> void sort(Vector<T>& a) // bubblesort
{
   int i, j;
   for(i=0;i<a.get_dim()-1;i++)
     for(j=a.get_dim()-1;i<j;j--)
       if(a[j] < a[j-1])
       {
       T temp = a[j];  a[j] = a[j-1];  a[j-1] = temp;
       }
}

void main(void)
{
   char *who = NULL;   who = new char[10];
   Vector<Person> V(3);
   Person a("Peter");  Person b("Rudi");  Person c("Ralf");
   V[0] = a; V[1] = b; V[2] = c;
   cout << " Is name in data base ? " << endl;
   cout << " name: ";
   cin >> who;
   Person e(who);
   find(V,e);
   delete [] who;
   sort(V);
   V.print();

   Vector<int> W(3);
   W[0] = 3; W[1] = 1; W[2] = 9;
   sort(W);
   W.print();
}
```

7.9 Student Names and Student Numbers

Given a set of names of students. Each student has a student number, say The names of the students are in lexicographical order. They are listed in the array `name`. The students numbers are listed in the array `number`. The input of the name of the student should display the student's number. If the name is not in the list the output should be `name not in list`.

The function `int lookup()` in the program `student1.cpp` takes the name of the student. It then finds the place (number) in the array, where the name is placed. This integer number is returned. If the name is not in the list -1 is returned. The integer number is now inserted into `number[]`. This yields the student number of the corresponding name.

```
// student1.cpp

#include <iostream.h>
#include <string.h>

char *name[] = { "Muller", "Smith", "Solms", "Steeb" };
char *number[] = { "101", "102", "103", "104" };

int lookup(char *str)
{
  int i;
  for(i=0; i<4;i++)
  {
  int result = strcmp(str,name[i]);
  if(result == 0)
  return i;
  }
  return -1;
}

void main()
{
  char* word = new char[10];
  cout << "enter student name: ";
  cin.getline(word,10);
  int x = lookup(word);
  if(x != -1) cout << "Student's number = " << number[x] << "\n";
  if(x == -1) cout << "name is not found on the list ";
}
```

In a second program student2.cpp we also give the marks of the students. Thus we have an array marks[ ]. Then the input of the name of the student will display the student's name and marks.

```
// student2.cpp

#include <iostream.h>
#include <string.h>

char* name[] = { "Muller", "Smith", "Solms", "Steeb" };

char* number[] = { "101", "102", "103", "104" };

char* marks[] = { "50%", "74%", "100%", "100%" };

int lookup(char *str)
{
  int i;
  for(i=0; i<4; i++)
  {
  int result = strcmp(str,name[i]);
  if(result == 0)
  return i;
  }
  return -1;
}

int main()
{
  char* word = new char[10];
  cout << "enter student name: ";
  cin.getline(word,10);
  int x = lookup(word);
  if(x != -1)
  {
  cout << "Student's number = " << number[x] << "\n";
  cout << "The marks are = " << marks[x] << "\n";
  }

  if(x == -1) cout << "name is not found on the list ";

  return 0;
}
```

7.10 Search for Names in a List

Given a list of names, for example

```
"Bob", "Walter", "Catherine", "Helen", "Sue", "Willi-Hans"
```

The program ask for a name to be entered. It then checkes whether the name is in the list or not. The function `strcmp()` is used to compare two strings. The `strcmp()` function compares two strings (in our case two names in alphabetical order, see ASCII code) and returns an integer value based on the comparison. If the string1 is on the righthand side within the parentheses and string2 is on the left then the integer value returned will have the following meanings:

(1) less than zero (string1 less than string2 with respect to alphabetical order)

(2) zero (string1 identical to string2 with respect to alphabetical order)

(3) greater than zero (string1 greater than string2 with respect to alphabetical order)

`char list[MAX][LEN]` is declared static. This means it retains its previous values each time function `main` is invoked.

```
// complist.cpp

#include <iostream.h>
#include <stdio.h>      // for gets()
#include <string.h>     // for strcmp()

void main()
{
   const int MAX = 6;   const int LEN = 30;
   int j;
   int enter = 0;
   char name[LEN];
   static char list[MAX][LEN] =
   { "Bob", "Walter", "Catherine", "Helen", "Sue", "Willi-Hans" };
   cout << " Enter name: ";
   gets(name);
   for(j=0;j<MAX;j++)
   if(strcmp(&list[j][0],name)==0)
     enter = 1;
   if(enter==1) cout << " Name is on the list ";
   else  cout << " Name is not on the list ";
}
```

7.11 Names of Employees in a Table

The following program finds a name in a table when the corresponding number (employee number) is given and vice versa. We apply the standard template library.

```
// table.cpp

#include <iostream>
#include <string>
#include <map>

using namespace std;

int main()
{
   map<int,string> m;

   m.insert(pair<int,string>(101,"Simpson"));
   m.insert(pair<int,string>(102,"Singer"));
   m.insert(pair<int,string>(103,"Carter"));
   m.insert(pair<int,string>(104,"Thompson"));
   m.insert(pair<int,string>(105,"Ulrich"));

   int stnumber;

   cout << "enter the students number: ";
   cin >> stnumber;

   map<int,string>::iterator p;

   p = m.find(stnumber);
   if(p != m.end())
   cout << p -> second;
   else
   cout << "not a student number";
   cout << endl;

   map<string,int> n;

   n.insert(pair<string,int>("Simpson",101));
   n.insert(pair<string,int>("Singer",102));
   n.insert(pair<string,int>("Carter",103));
   n.insert(pair<string,int>("Thompson",104));
```

```
   n.insert(pair<string,int>("Ulrich",105));

   string name;

   cout << "enter the students name: ";
   cin >> name;

   map<string,int>::iterator q;

   q = n.find(name);
   if(q != n.end())
   cout << q -> second;
   else
   cout << "not a student name";

   return 0;
}
```

7.12 The Divide-and-Conquer Methodology

The following program finds the maximum and minimum elements in an array of integers. The number of array elements must be 2^n, where n is an integer. We apply the divide-and-conquer algorithm. Such an algorithm divides the problem into smaller pieces, solves each piece either directly or by recursion, and recombines the solution of the parts into the solution of the whole.

```
// minmax.cpp

#include <iostream.h>

void maxmin(int *a, int n, int *minp, int *maxp)
{
   int min1, max1, min2, max2;
   if(n == 2)
      if(a[0] < a[1]) { *minp = a[0]; *maxp = a[1]; }
      else { *minp = a[1]; *maxp = a[0]; }
      else {
      maxmin(a, n/2, &min1, &max1);
      maxmin(a + n/2, n/2, &min2, &max2);
      if(min1 < min2)
      *minp = min1;
      else
      *minp = min2;
      if(max1 < max2)
      *maxp = max2;
      else
      *maxp = max1;
      }
}

int main()
{
   const int n = 8;
   int a[n] = { 11, 5, 7, 9, -11, 8, 0, 17 };
   int min, max;
   maxmin(a,n,&min,&max);
   cout << " minimum is: " << min << "\n";
   cout << " maximum is: " << max << "\n";
   return 0;
}
```

7.13 Sorting Vectors

Given a set of two-dimensional vectors, say (4,5) (1,1) (5,3) (0,8). The following program ptsort.cpp sorts the vectors with respect to their lengths (norms). If $\mathbf{x} = (x_1, x_2)$ is a vector, the length (Euclidian norm) is given by

$$\|\mathbf{x}\| := \sqrt{x_1^2 + x_2^2}.$$

For the sorting of the vectors it is not necessary to take the square root of $x_1^2 + x_2^2$, since

$$\text{if} \quad \sqrt{y_1^2 + y_2^2} \le \sqrt{x_1^2 + x_1^2} \quad \text{then} \quad y_1^2 + y_2^2 \le x_1^2 + x_2^2.$$

Thus the sorted vectors are (1,1) (5,3) (4,5) (0,8). We use the function qsort in the program. This function is included in the header file <string.h>.

```
// ptsort.cpp
// we sort two-dimensional vectors according to their length

#include <stdio.h>
#include <stdlib.h>  // for qsort

// Declaration of class Point
class Point
{
   private:
      int m, n;
   public:
      Point();
      Point(int,int);
      friend Ptarray;
};

// Definition of class Point
Point::Point() { m = 0;  n = 0; }

Point::Point(int mi,int ni) { m = mi;  n = ni; }

// Declaration of class Ptarray
class Ptarray
{
   private:
      Point* ptr;
      int nptr;
      static int compare(const void*,const void*);
   public:
```

```
      Ptarray(int);
      ~Ptarray();
      void set(int,int,int);
      void sort();
      void print();
};

Ptarray::Ptarray(int np) { ptr = new Point[nptr = np]; }

Ptarray::~Ptarray() { delete [] ptr; }

int Ptarray::compare(const void* a,const void* b)
{
   Point* pa = (Point*)a;
   Point* pb = (Point*)b;
   return pa->m*pa->m + pa->n*pa->n - pb->m*pb->m - pb->n*pb->n;
}

void Ptarray::set(int k,int m,int n) { ptr[k].m = m;  ptr[k].n = n; }

void Ptarray::sort()
{
   qsort(ptr,nptr,sizeof(Point),Ptarray::compare);
}

void Ptarray::print()
{
   printf("The sorted points are:\n\n");
   int i;
   for(i=0; i <nptr; i++)
   printf("%d: (%02d, %02d)\n",i,ptr[i].m,ptr[i].n);
}

void main()
{
   Ptarray pa(4);   // four vectors
   pa.set(0,4,5);   // vector number 0 with components (4,5)
   pa.set(1,1,1);   // vector number 1 with components (1,1)
   pa.set(2,5,3);   // vector number 2 with components (5,3)
   pa.set(3,0,8);   // vector number 3 with components (0,8)
   pa.sort();
   pa.print();
}
```

Chapter 8

Useful Classes

8.1 Introduction

In this chapter we introduce a collection of useful classes. These classes can then be included in other C++ programs in order to be used in this class. They define an abstract data type (for example string, vector, matrix, linked list). An abstract data type is a collection if data objects together with a set of operations.

First we develop several string classes. Since a string type is lacking in C and C++ such a class is very helpful. Such a string class could then be used in a name sorting program. Next we discuss several vector classes. A template vector class is also described. In section 8.4 a matrix class with all the standard matrix operations is implemented. A template class is used. A linked list is a sequence of zero or more nodes, together with a set of operations. Each node contains two fields, an `item` field that holds information and a `next` field that holds a pointer to the next node in the list. A linked list is a finite sequence of nodes that that each has a pointer field containing a pointer to the next node. In section 8.5 linked lists are introduced. We also describe a circular linked list, a recursive linked list, and a intrusive linked list. The generalized linked list (i.e. a list which can contain lists itself) is the basic abstract data type in some languages. The primary example would be LISP, which stands for LIST PROCESSING. We also give an implementation of a generalized list. Binary trees are considered in section 8.6. A tree is composed of a number of tree nodes. Each node can have one or more children. A binary tree is restricted to having at most two children. Section 8.7 provides a set class for dealing with finite sets. The union and intersection of finite sets are implemented. A date class is described in section 8.8.

8.2 String Class

8.2.1 Introduction

A string type is lacking in traditional C. Strings in C are represented as a pointer to a `char` and are manipulated accordingly. This means strings can be considered as arrays of type `char`. In this representation, the end-of-string is denoted by the null character `'\0'`. This convention has an important drawback, in that many basic string manipulations are proportional to string length. When the string length is known, the efficiency of operations on strings can be improved. We give several string classes, where the strings are considered as abstract data types.

The first program `string1.cpp` gives the implementation of a simple string class. It shows that more than one constructor can be defined. The member function `concat` concatenates two strings. The member function `display` displays the strings on the screen. We use two library functions: `strlen` and `strcpy`. The private data members are a pointer to `char` (for the string) and a variable `length` in which the current string length is stored. The constructors allocate the memory dynamically from free store. If we allocate memory (using the `new` operator) when we create an object, we should deallocate the memory when the object is no longer needed. The destructor gives back the memory obtained when the object was created. It also includes a copy constructor and the assignment operator `=` is overloaded. Furthermore the operator `+` is overloaded in order to concatenate two strings. The copy constructor

```
String(const String&);
```

is a constructor that copies its arguments into a new object. The default copy constructor, which is provided automatically by the compiler for every object, performs a member-by-member copy. This is similar to what the assignment operator does. The difference is that the copy constructor also creates a new object.

The program `string2.cpp` gives a comprehensive implementation of a string class. The programs `string1.cpp` is included as subcase.

Next we introduce the string class of the standard library. All newer C++ compilers support this string class. We list all the constructors and methods in the class. Then we give an example.

8.2.2 A Simple String Class

```
// string1.cpp

#include <iostream.h>
#include <string.h>    // for strlen, strcpy, strcat

     // Declaration of class String
class String
{
   private:
      char *s;                    // pointer to string contents
      int length;                 // length of string in characters
   public:
      String(char*);     // constructor using existing string
      String(int);                // creates default empty string
      String(const String&);  // copy constructor
      ~String();                  // destructor
      String operator + (const String&);    // concatenate
      String& operator = (const String&);   // assignment operator
      void concat(const String&,const String&);
      void display();
};

    // Definition of class String
String::String(char *text)
{
   length = strlen(text);
   s = new char[length+1];
   strcpy(s,text);
}

String::String(int size)
{
   length = size;     // get length of string
   s = new char[length+1];
   *s = '\0';
}

String::String(const String& other)
{
   length = other.length;         // length of other string
   s = new char[length+1];        // allocate the memory
   cout << "copy constructor invoked " << endl;
   strcpy(s,other.s);            // copy the text
```

```
}

String::~String()
{
  delete s;
}

String String::operator + (const String& arg)
{
  String temp(length + arg.length);
  strcpy(temp.s,s);
  strcat(temp.s,arg.s);
  return temp;
}

String& String::operator = (const String &arg)
{
  if(this == &arg)
  { return *this; }
  delete [] s;
  s = new char[strlen(arg.s) + 1];
  strcpy(s,arg.s);
  cout << "assignment invoked " << endl;
  return *this;
}

void String::concat(const String& s1, const String& s2)
{
  delete s;                                  // forget current string
  length = strlen(s1.s) + strlen(s2.s);
  s = new char[length];                      // allocate the memory
  strcpy(s,s1.s);                            // copy first into space
  strcat(s,s2.s);                            // add the second
}

void String::display()
{
  cout << s;
}

int main()
{
  cout << endl;
  char *str = "to be ";
  String a(str);
```

```
   String b("or not ");
   String c("to be this is the question");
   String quote1("");
   String quote2("");
   quote1.concat(a,b);
   quote1.display();
   cout << endl;
   quote2.concat(quote1,c);
   quote2.display();
   cout << endl << endl;

   String u("much ado ");
   String v("about nothing by William Shakespeare");
   String w("");
   w = u + v;
   w.display();
   cout << endl << endl;

   return 0;
}
```

The output is

```
copy constructor invoked
assignment invoked
much ado about nothing by William Shakespeare
```

8.2.3 Advanced String Classes

In the following we give a more advanced string class. The operators

`=    ()    []    ==    <    <=    <<    <<`

are overloaded.

```
// string2.cpp

#include <iostream.h>
#include <string.h>    // for strlen, strcpy, strcat, strncpy
#include <assert.h>    // for assert

      // Declaration of class String
class String
{
   private:
      char* s;
      int length;
   public:
      String(const char*);
      String(unsigned);
      String(unsigned,char);
      String(const String&);     // copy constructor
      ~String();                 // destructor
      String& operator = (const String&);
      String operator () (unsigned, unsigned) const;
      String operator () (unsigned) const;
      char& operator [] (unsigned);
      operator char* () const;
      friend String operator + (const String&,const String&);
      friend int operator == (const String&,const String&);
      friend int operator != (const String&,const String&);
      friend int operator < (const String&,const String&);
      friend int operator <= (const String&,String&);
      friend ostream& operator << (ostream&,const String&);
      friend istream& operator >> (istream&,String&);
      void display();
      void read(unsigned);
      String swap_char(unsigned,unsigned);
      int replace(String,String);
      String reverse() const;
      void fill(char);
};
```

```
String::String(const char* p)
{
   length = strlen(p);
   s = new char[length+1];
   strcpy(s,p);
}

String::String(unsigned len)
{
   length = len;
   s = new char[length+1];
   s[length] = '\0';
}

String::String(unsigned len,char tofill)
{
   length = len;
   s = new char[length+1];
   fill(tofill);
   s[length] = '\0';
}

String::String(const String& toCopy)
{
   length = toCopy.length;
   s = new char[length+1];
   strcpy(s,toCopy.s);
}

String::~String()
{
   delete s;
}

String& String::operator = (const String& toSet)
{
   if(length != toSet.length)
   {
      delete s;
      length = toSet.length;
      s = new char[length+1];
   }
   strcpy(s,toSet.s);
   return *this;
```

```
}

String String::operator () (unsigned start,unsigned end) const
{
   assert(start > 0 && end <= length && start <= end);
   int newlength = end - start + 1;
   String S(newlength);
   strncpy(S.s,s + start - 1, newlength);
   return S;
}

String String::operator() (unsigned end) const
{
   assert(end > 0 && end <= length);
   return (*this)(1,end);
}

char& String::operator [] (unsigned i)
{
   assert(i > 0 && i <= length);
   return s[i-1];
}

String::operator char* () const
{
   char* p = new char[length];
   strcpy(p, s);
   return p;
}

String operator + (const String& s1,const String& s2)
{
   String S(s1.length + s2.length);
   strcpy(S.s,s1.s);
   strcat(S.s,s2.s);
   return S;
}

int operator == (const String& s1,const String& s2)
{
   return !strcmp(s1.s,s2.s);
}

int operator != (const String& s1,const String& s2)
{
```

```
   return !(s1 == s2);
}

int operator < (const String& s1,const String& s2)
{
   return strcmp(s1.s,s2.s);
}

int operator <= (const String& s1,const String& s2)
{
   return (s1 < s2 || s1 == s2);
}

ostream& operator << (ostream& out,const String& S)
{
   out << S.s;
   return out;
}

istream& operator >> (istream& in,String& S)
{
   in >> S.s;
   S.length = strlen(S.s);
   return in;
}

void String::display()
{
   cout << s;
}

void String::read(unsigned i)
{
   length = i;
   delete s;
   s = new char[length+1];
   cin >> s;
}

String String::swap_char(unsigned n,unsigned m)
{
   String S(length);
   char temp;
   strcpy(S.s, s);
   temp = S[n];
```

```
   S[n] = S[m];
   S[m] = temp;
   return S;
}

int String::replace(String sub,String new_sub)
{
   char *temp, *ptr, *base;
   int len;
   if((sub.s == '\0') || (new_sub.s == '\0'))
      return 0;
   else {
      temp = new char[length];
      len = sub.length;
      base = s;
      while((base = strstr(base, sub.s)) != NULL)
      {
      ptr = base + len;
      strncpy(temp, ptr, strlen(ptr)+1);
      strcpy(base, new_sub.s);
      strcpy(base + new_sub.length,temp);
      }
      if(ptr == s)
      cout << "sorry substring cannot be replaced.\n";
      else
      cout << "The new string is " << s << endl;
      delete temp;
      return 1;
    }
}

String String::reverse() const
{
   String S(length);
   int i;
   for(i=0; i<length; i++)
      S.s[i] = s[length-i-1];
   return S;
}

void String::fill(char tofill)
{
   int i;
   for(i=0; i < length; s[i++] = tofill);
}
```

```
int main()
{
   String S1("Johannesburg");
   cout << "S1 = " << S1 << endl;  // Johannesburg
   S1.display();                   // Johannesburg
   cout << endl;

   String S2(3);
   cout << "S2 = " << S2 << endl;

   String S3(6,'$');
   cout << "S3 = " << S3 << endl;  // $$$$$$

   String S4(S1);
   cout << "S4 = " << S4 << endl;

   S2 = S4;
   cout << "S2 = " << S2 << endl;  // Johannesburg

   S4.fill('+');
   cout << "S4 = " << S4 << endl;  //  ++++++++++++
   cout << "S2 = " << S2 << endl;  //  Johannesburg

   cout << "reverse of S1: " << S1.reverse() << endl; // grubsennahoJ
   S1.reverse().display();                            // grubsennaboJ
   cout << endl;

   String S5(30);
   S5 = S1 + S1.reverse();
   cout << "S5 = " << S5 << endl;    // JohannesburggrubsennahoJ


   cout << "S1 == S2 --> " << (S1 == S2) << endl;   // 1
   cout << "S1 == S4 --> " << (S1 == S4) << endl;   // 0

   cout << "S1.swap_char(2,6): " << S1.swap_char(2,6); // Jnhanoesburg
   cout << endl;

   return 0;
}
```

8.2.4 Standard Library and String Class

Most applications that must send textual information to a human reader, or process input received from a user interface, will employ character values. However, the basic C++ language does not provide a string type as a primitive, instead it builds rudimentary strings out of pointers and arrays of characters, in a fashion C++ inherited from its parent language, C. For this reason, a string class is a logical candidate for development as a library class.

A string is a sequence of characters. In fact, nearly all the operations associated with arrays or vectors can be applied equally well to strings. However, a string is also a much more abstract entity, in addition to simple vector operators, the string datatype provides a number of useful and powerful high-level operations.

One confusing aspect of the `string` facility is the presence of two different include files. The statement

```
#include <string.h>
```

includes a file that declares a number of functions useful in conjunction with older, C-style character values, for example `strlen`, `strcpy`, `strcmp` etc.. To use the `string` data type, the programmer must include the `string` file (without the `.h` extension).

```
#include <string>
```

The Draft C++ Standard provides a library string class, to make string manipulation easier and less error-prone. The string class does not belong to STL but it is supported by all newer C++ compilers.

To define string variables, we have to use the following `#include` line:

```
#include <string>
```

Now we can do the following assignments

```
string s, t;
s = "ABC";
t = "DEFGH";
s = t;        // s = "DEFGH"
s = 'A';      // s = "A"
```

The following table summarizes the constructors and member functions (methods) of the string class.

Table. Operations for the string data type

```
Constructors
string s;           default constructor
string s("text");   initialized with literal string
string s(aString);  copy constructor

Character Access
s[i]                subscript access
s.substr(pos,len)   return substring starting at position
                    of given length
s.c_str()           return a C-style character pointer

Length
s.length()          number of characters in string
s.resize(int,char)  change size of string, padding with char
s.empty()           true if string has no characters

Assignment and Catenation
s = s2;  assignment of string
s += s2; append second string to end of first
s + s2   new string containing s followed by s2

Iterators
string::iterator t  declaration of new iterator
s.begin()           starting iterator
s.end()             past-the-end iterator

Insertion, Removal, Replacement
s.insert(pos,str)             insert string after given position
s.remove(start,length)        remove length characters after start
s.replace(start,length,str)   insert string, replacing
                              indicated characters

Comparisons
s == s2  comparision for equality
s != s2  comparisons for inequality
s < s2   comparisons for relation
s > s2   comparisons for relation
s <= s2  comparisons for relation
s >= s2  comparisons for relation
```

```
Searching Operations
s.find(str)                    find start of argument string
                               in receiver string
s.find(str,pos)                find with explicit starting position
s.find_first_of(str,pos)       first position of first character
                               from argument
s.find_first_not_of(str,pos)   first character not from argument

Input / Output Operations
stream << str                  output string on stream
stream >> str                  read word from stream
getline(stream,str,char)       read line of input from stream
```

Notice that the comparison operators

```
==  !=
```

as well as the relational operators

```
<  <=  >  >=
```

compare the contents of the two string arguments, not their addresses in memory, as is the case with character pointers.

The following program shows an application of the string class.

```
// mystring.cpp

#include <iostream>
#include <string>

using namespace std;

int main()
{
   string str1("The string class provides C++");
   string str2(" with string handling.");
   string str3;

   // assignment
   str3 = str1;
   cout << str3 << endl; // =>

   // comparison
   int result = (str1 != str2);
```

```
   cout << "result = " << result << endl; // =>

   // concatenate two strings
   // use overloaded +
   str3 = str1 + str2;
   cout << str3 << endl;

   // length of string
   unsigned long a = str3.length();
   cout << "length of string str3 = " << a;

   bool b = str1.empty();
   cout << "b = " << b << endl;

   string str4("Good Morning Egoli");
   string str5("XYZ1234");

   cout << "Initial string:\n";
   cout << "str4: " << str4 << endl;
   cout << "str5: " << str5 << endl;

   // method insert
   cout << "insert str5 into str4:\n";
   str4.insert(5,str5);
   cout << str4 << "\n\n";

   // method erase
   cout << "remove 7 characters from str4:\n";
   str4.erase(5,7);
   cout << str4 << "\n\n";  // =>

   // method replace
   cout << "replace 2 characters in str4 with str5:\n";
   str4.replace(5,2,str5);
   cout << str4 << endl;  // =>

   return 0;
}
```

8.3 Vector Class

8.3.1 Introduction

Arrays are common data structures that are supported by C and C++. However, C and C++ arrays have some weaknesses. They are effectively treated as typed pointers. Consequently no bound checking is performed when using the built in array types. On the other hand, if we develop an array based class (e.g. vector or matrix class) then we can incorporate bound checking into these classes. Furthermore, we are then able to define mathematical operators for these data types (for example vector addition or matrix multiplication). It is thus useful to introduce a class as an abstract data type for an array. Here we consider one-dimensional arrays. Thus we call the class Vector. The vector class is an abstract data type. An abstract data type is defined as a collection of data objects together with a set of operations (member functions of the class).

Dynamic memory data types are very useful. Most array based data types (e.g. vectors and matrices) should generally be defined dynamically so that the memory required adjusts itself to the need of the application. Furthermore, these data types can require huge amounts of memory and it may be advisable to release the memory as soon as it is no longer required. Other very common dynamic data types are linked lists and trees.

For dynamic data types it is essential to write our own constructors and in particular the copy constructor. Furthermore, one has to define a destructor which releases the dynamically allocated memory as soon as the object goes out of scope and an assignment operator which frees and allocates resources as necessary.

In this section we consider vector classes. In subsection 8.3.2 we give a simple vector class. A copy constructor is introduced and the assignment operator = is overloaded. The operator `[ ]` is overloaded to access the elements of the vector. In section 8.3.3 we describe in detail the construction of an advanced template vector class.

8.3.2 A Simple Vector Class

In this section we give a simple vector class. The constructor

```
Vector(int)
```

allows the user to build dynamically allocated arrays. In the `main` function the line

```
Vector v(20);
```

creates an vector object `v` to hold 20 `double`.

A copy constructor is implemented. Moreover the destructor `~Vector` is implemented. The subscript operator `[ ]` is overloaded. An overloaded subscript operator can have any return type and any argument list type. The assignment operator `=` is also overloaded. The `operator =` function returns a reference to `Vector` and has an explicit argument of type reference to `Vector`. We could also have added a constructor

```
Vector(const double a[],int)
```

for the initialization by array. The implementation is

```
Vector::Vector(const double a[],int n)
{
   if(n <= 0)
   {
   cerr << "illegal vector size: " << n << endl;
   exit(1);
   }
   dim = n;
   array = new double[dim];
   int i;
   for(i = 0; i < dim; i++)
      array[i] = a[i];
}
```

The implementation of vector1.cpp is

```
// vector1.cpp

#include <iostream.h>

     // Declaration of class Vector
class Vector
{
   protected:
      int dim;
      double *array;
   public:
      Vector(int);
      Vector(const Vector&);
      ~Vector();
      Vector& Vector::operator = (const Vector&);
      Vector operator + (Vector&);
      double& operator [] (int);
      int Dim();
      void print();
};

  // Definition of class Vector
Vector::Vector(int n)
{
   dim = n;
   if(dim > 0)
   array = new double[dim];
   else
   {
   dim = 0; array = NULL;
   }
}

Vector::Vector(const Vector& v)
{
   dim = v.dim;
   array = new double[dim];
   int i;
   for(i = 0; i < dim; i++)
   array[i] = v.array[i];
   cout << "copy constructor invoked " << endl;
}
```

```
Vector::~Vector()
{
   if(array) delete [] array;
}

Vector& Vector::operator = (const Vector& v)
{
   int s = (dim < v.dim) ? dim : v.dim;
   if(v.dim != dim)
   cerr << "copying different size arrays "
        << dim << " and " << v.dim << endl;
   int i;
   for(i=0; i < s; i++)
      array[i] = v.array[i];
   return *this;
}

double& Vector::operator [] (int pos)
{
   return array[pos];
}

Vector Vector::operator + (Vector& arg)
{
   Vector res(arg);
   int i;
   for(i=0; i < dim; i++)
   {
   res.array[i] = array[i] + arg.array[i];
   }
   return res;
}

int Vector::Dim()
{
   return dim;
}

void Vector::print()
{
   int i;
   cout << "(";
   for(i=0; i < (dim-1); i++)
   cout << array[i] << ",";
   cout << array[dim-1] << ")" << endl;
```

```
}

int main()
{
   int m = 3;
   Vector v1(m);
   Vector v2(m);
   v1[0] = 1.2;  v1[1] = 0.3;  v1[2] = 2.0;
   v2[0] = -7.1;  v2[1] = 0.3;  v2[2] = 4.1;

   v1.print();
   v2.print();

   Vector v3(m);
   v3 = v1 + v2;
   v3.print();

   return 0;
}
```

The output is

```
(1.2,0.3,2)
(-7.1,0.3,4.1)
copy constructor invoked
copy constructor invoked
(-5.9,0.6,6.1)
```

8.3.3 Construction of an Advanced Vector Class

In this section we describe the construction of an advanced template vector class. Concepts which will be illustrated are class templates, classes which own other classes (have other classes as members), friend functions, and the need and usefulness of conditional compilation. These concepts are introduced through developing a vector class and a matrix class. In order to demonstrate the power and convenience of these classes, we develop a a library function which calculates the quantities important for these classes.

We view a vector as an abstract data type consisting of a one-dimensional array together with the operations which can be performed on that array. We define the vector on a template so that we have one code for vectors of various data types (e.g. `int`, `float`, `double`, and `complex`) as well as user defined types (e.g. rational numbers, strings, records of employees, ...). Our aim for this section is to develop a vector class that can be used as follows:

```
vector<double> v1(n);  // declaring vector of n double
                       // precision numbers.
vector<double> v2(v1); // declaring vector v2 as copy of v1.

vector<int> vint(7);   // declaring an integer vector of 7 elements.

vector<double> v3 = v1+v2; // vector addition and assignment.

double dotprod = v1%v2;    // dot product between two vectors.

v3.normalize();            // normalize vector v3.

v3[3] = v1[1]*v2[2];       // accessing vector elements

cout << << "vector v3 = " << v3 << endl; // stream access
```

We are thus able to create vector objects of various types, do vector addition, vector assignment, calculate the dot product between two vectors, normalize vectors, access the individual elements and push a vector onto the output stream. The implementation details of the vector class will be hidden from the person who uses it as if it were a built-in data type.

Instead of declaring the data fields private, we declare them protected in order to give the derived classes (which we will develop in the next chapter) access to these fields. Hence, class members declared private can only be accessed from within the class, members declared protected can be accessed from within the class as well as within derived classes and public members can be accessed by users of the class.

The protected data fields for the class are a pointer to the template parameter `T` which will point to the memory area of the vector and an unsigned integer `lngth`:

```
template <class T> class vector
{
  protected:
    T* v;
    unsigned lngth;
  public:
    ...
}
```

The vector class is defined on a template and the data type of a vector object is determined when declaring a vector. For example the following statement declares a vector of three rational numbers

```
vector<rational> ratvec(3);
```

The user-defined rational class can be replaced by any other built-in or user-defined data type (e.g. by `float`).

We need one constructor `vector(n)` which reserves us memory for `n` elements:

```
template <class T>
vector<T>::vector(const unsigned n) // constructor for vector
{                                   // of n elements
  v = new T[n];
  assert (v);
  lngth = n;
}
```

The macro `assert`, which is part of the ANSI standard for C, checks whether the `new` statement failed (i.e. when the heap is exhausted) by checking whether `v` is equal to `0`. If it did fail it will print the error message:

```
Assertion failed: (v), file vector.cpp, line 93
```

and it will call `exit()` to flush the output streams, close the files and abort the program.

If the `new` statement was successful, we reserve memory for `n` array elements, set the pointer `v` to the start of that memory area and set the data field `lngth` equal to `n`. The default constructor is required for defining arrays of vectors (we shall define our matrix class on arrays of vectors). It simply sets the pointer `v` to `NULL` and `lngth` to `0`.

Should one omit to define an assignment operator and a copy constructor, the C++ compiler will write them. However, one should bear in mind that the code produced by the compiler simply makes a byte for byte copy of the data members. In the case where the class allocates memory dynamically, we have to write the assignment operator = and copy constructor. We avoid code duplication by letting the copy constructor make use of the assignment operator of the class. When assigning `v1` to `v2` (`v2=v1`), we first release the memory area for the old vector `v2`. Then sufficient memory is reserved for `v2` for it to hold a copy of `v1` and finally the elements of `v2` are set equal to those of `v1`:

```
template <class T>
vector<T>& vector<T>::operator=(const vector<T>& v1)
{
  if(this != &v1)  // check for self assignment
  {                 // (dangerous because of delete []v)
    if(v)
    delete []v;
    resize(v1.length());
    unsigned i;
    for(i=0; i<lngth; i++)
      v[i] = v1[i];
  }
  return *this;
}
```

It is critical to check for self-assignment. Self assignment can be more subtle than the simple `v1=v1`. For example, `v2` could be a reference variable which has been set to `v1` and in such a case `v2=v1` would be a camouflaged self-assignment. The check for self-assignment is not only recommended for reasons of efficiency. Observe that the memory holding the only copy of the data fields (the elements of the vector) would be released before the actual copying is done.

Note again that the assignment operator is defined such that it returns a reference to the object itself, thereby allowing constructs like

```
v1 = v2 = v3;
```

The copy constructor makes direct use of the assignment operator defined above:

```
template <class T> inline
vector<T>::vector(const vector<T>& v1)   // copy constructor
{ v=NULL; *this = v1; } // see assignment operator
```

Note that the memory management and copying of data fields is all done by the assignment operator. Destructors are called automatically when an object goes out of scope. Hence if we define a vector locally in a function that vector's destructor will be called (and hence the memory reserved for the vector will be released) as soon as the function is exited. They can also be called manually via the `delete` operator:

```
delete vector;
```

Constructors are essential for classes which allocate memory dynamically. Similarly to constructors, destructors carry the name of the class, but with a ~ inserted in front of the name. Hence, `~vector()` is the destructor for the vector class. Destructors have no arguments and no return values. The body for destructor of the vector class is very simple

```
template <class T> vector<T>::~vector()
{ if(v) delete []v; }
```

It simply checks whether there has been memory allocated by checking whether the pointer `v` is not the `NULL`-pointer and, if so, it releases the memory via the standard `delete` statement and sets the pointer `v` equal to the `NULL`-pointer.

The member functions of the vector class are straightforward. Arrays of objects are declared using the default constructor which declares a vector of zero length. This prompts us to define a public method

```
resize(newlength)
```

which allows us to resize each vector in the array. The function `resize` will be used by the matrix class which builds a matrix object as an array of vectors.

The member function `norm()` returns the norm of the vector and `normalize()` normalizes the vector object (it in turn uses `norm()`). The function `unit(i)` sets the vector object to the unit vector $\mathbf{e}_i$, `fill(x)` sets all elements of the vector equal to `x` and `swap(i,j)` swaps elements `i` and `j`.

In standard C++ arrays the elements are accessed via the subscription operator `[]`. For example, in

```
float* x[3];
cout << "2nd element of vector x = " << x[1] << endl;
```

`x[1]` refers to the 2nd element of the vector `x`. To remain with this tradition of C and C++ we overload the operator `[]` to gain access to the vector elements. This can be done as follows:

```
template <class T> class vector
{
     ...
```

```
  public:
    ...
    T& operator [] (const unsigned i) const
   { assert(i<lngth); return v[i]; }
    ...
}
```

Since the subscript operator should leave the object unaltered, it is defined as a constant method. The assert macro checks for out of bounds. This is very useful while developing a code, but the performance penalty for typical vector and matrix applications is around 20%, which is very undesirable. In the case of the `assert()` macro one simply adds

```
#define NDEBUG
```

above the

```
#include <assert.h>
```

compiler directive. This will effectively comment out the `assert` statement. If one does not use `assert` for error checking, one can still achieve conditional compilation of bound-checking code as follows:

```
#define boundcheck 1  // set to 1/0 for bound checking on/off
...
#if boundcheck
  template <class T>
  T& vector<T>::operator::(const unsigned i) const
  {
    if(i>=lngth)
    {
      cerr << "ERROR: out of bounds in vector::[]" << endl;
      exit(1);
    }
#else
  template <class T> inline
  T& vector<T>::operator::(const unsigned i) const
  {
#endif
return v[i];
}
```

The remaining operators are straightforward and can be studied from the program listing. We have used `*` for element by element multiplication and `%` for the vector dot product. This is a good choice since the operator `%` is a binary operator with the same predefined precedence level as multiplication and division. Recall that one cannot alter the order of precedence of the operators, nor their syntax. The precedence hierarchy and the syntax of C++ operators is summarized in chapter 0.

The dot product is defined as follows:

```
template <class T>
T vector<T>::operator%(const vector<T>& v1) const
{
  T result = 0.0;

  assert(v1.length() == lngth);
  unsigned i;
  for(i=0; i<=lngth-1; ++i)
    result += v1[i]*v[i];
  return result;
}
```

Again we have used the `assert` macro to check that the two vectors in the dot product are of the same length. The result of the dot product is of type `T` defined by the template. Note that it is returned by value and not by reference. One should never return a reference to a local variable. A local variable is destructed as soon as it goes out of scope and we would be returning a reference to an object that no longer exists. The dot product would be used as follows:

```
vector<double> v1(3), v2(3);
...
double result = v1%v2;
```

The multiplication operator, on the other hand, returns a vector whose elements are the product of the equivalent elements in `v1` and `v2`. The implementation details are given in the following listing:

```
template <class T>
vector<T> vector<T>::operator * (const vector<T>& v1) const
{
  vector<T> v2(v1);

  assert(v1.length() == lngth);
  unsigned i;
  for(i=0;i<lngth;i++)
    v2[i] *= v[i];
  return v2;
}
```

The return value is a vector which has to be passed by value.

For convenience, C and C++ have defined the operators *=, -=, +=, etc. For some classes it might be a good idea to continue to support these operators. In order to avoid code duplication, we define these operators in terms of the more basic operators (= and * in the case of *=)

```
template <class T> inline
void vector<T>::operator *= (const vector<T>& v1)
{
  *this = *this*v1;
}
```

Since multiplication has higher precedence than assignment, the object itself is first multiplied with another vector and the result is assigned to itself.

We also would like to multiply a vector of type T with a variable of type T (e.g. an integer vector with an integer variable). To be able to do this, we overload the multiplication operator as follows:

```
template <class T>
vector<T> vector<T>::operator * (const T& x) const
{
  vector<T> v1(lngth);
  unsigned i;
  for(i=0; i<=lngth-1; i++)
    v1[i] = v[i] * x;
  return v1;
}
```

The operator

```
vector<T>::operator * (const T& x) const
```

enables us to multiply a vector of type T with a variable of type T. Hence

```
vector<double> v1(3), v2(3);
...
double x = 4.1;

v2 = v1*x;  // no problem
v2 = x*v1;  // error: undefined operator
```

The cause for this problem is illustrated clearly by writing the two statements in functional form:

```
v2 = v1.operator*(x);
v2 = x.operator*(v1);
```

However, for floating point numbers, there is no operator defined that takes a vector as its parameter. In this case we define a global operator which takes a variable of type `T` as its left hand side argument and a vector as its right hand side argument:

```
template<class T>  // for x*vec;
vector<T> operator * (const T& x,const vector<T>& vec)
{
  return vec*x;
}
```

Note that `v*x = x*v` . Again we avoid code duplication by making use of the operator

```
vector<T>::operator * (const T& x) const
```

In certain cases one wants a function defined outside a certain class to have access to the private fields of that class. In such cases the function can be declared a friend of that class by including the function header in the class interface with the keyword friend inserted in front of the function header.

In many cases the output stream operator requires access to data fields that are declared private (or protected). In that case one can declare him a friend of the class by including the the keyword `friend` followed by the prototype declaration for the operator in the class interface:

```
template<class T>
ostream& operator << (ostream& os,const vector<T>& vec)
{
  unsigned n;
  os << "( ";
  for(n=0; n<vec.lngth; n++) // is friend of vector
    os << vec.v[n] << ' ';  // => access to protected field v
  return os << ")";
}
```

For the `vector` class it is not strictly necessary to declare the `<<` operator a friend, since we could have used the public member function `length()` and the public subscription operator `[]` of the class to retrieve the desired information.

8.3.4 Listing of Vector Class

```
// VECTOR.CPP
// Usage: vector<T> v(k); defines a vector of a built-in
// or user-defined data type, T, with +, -, * for element
// operations, % for dot product, normalize(), and other functions.
// The vector elements are accessed via vector[0],...,vector[length-1].
// Use #define NDEBUG to switch of bound checking and other
// checking options.

#include <iostream.h>
#include <assert.h>
#include <math.h>
#include <mem.h>

template<class T> class vector
{
  protected:
    T* v;
    unsigned lngth;
  public:
    // Constructors and Destructor
    vector();                    // default constructor
    vector(const vector& v1); // copy constructor
    vector(const unsigned l); // constructor for vector of length n
    ~vector();                   // destructor
    // Member functions
    void resize(const unsigned newlength); // resize vector
    void swap(const unsigned n1,const unsigned n2);
                                  // swap elements n1 and n2
    void fill(const T& x);        // set all elements = x
    void unit(const unsigned i); // sets v to unit vector e_i
    void normalize();             // normalize vector
    // Constant Member functions
    unsigned length() const; // return length of vector
    T maxel() const;            // returns largest element
    T maxabsel() const;        // returns largest abs(elem)
    T norm() const;        // returns sqrt(vec*vec)
    // Operators
    T& operator [] (const unsigned pos) const;
    vector<T>& operator =  (const vector<T>& v1);
    vector<T>  operator *  (const vector<T>& v1) const;
    vector<T>  operator +  (const vector<T>& v1) const;
    vector<T>  operator -  (const vector<T>& v1) const;
    T operator % (const vector<T>& v1) const;
```

```
    void operator *= (const vector<T>& v1);
    void operator += (const vector<T>& v1);
    void operator -= (const vector<T>& v1);
    vector<T> operator * (const T& x) const;
    vector<T> operator / (const T& x) const;
    vector<T> operator + (const T& x) const;
    vector<T> operator - (const T& x) const;
    void operator += (const T& x);
    void operator -= (const T& x);
    void operator *= (const T& x);
    void operator /= (const T& x);
    vector<T>& operator + (); // unary +
    vector<T>& operator - (); // unary -
    friend ostream& operator << (ostream& os,const vector<T>& vec);
}

// Globally defined operator for mixed algebra
template <class T>
vector<T> operator * (const T& x,const vector<T>& vec);

//                     Implementation section
template <class T> inline vector<T>::vector() : v(NULL), lngth(0) {};

template <class T> vector<T>::vector(const unsigned l)
{
  v = new T[l];
  assert(v);
  lngth = l;
}

template <class T> inline vector<T>::vector(const vector<T>& v1)
{ v = NULL; *this = v1; }

template <class T> vector<T>::~vector()  { if(v) delete []v; }

//                Bodies for Member Fuctions
template <class T>
void vector<T>::resize(const unsigned newlength)
{                               // resize vector to newlength
  delete v;
  lngth = 0;
  if(newlength > 0)
  {
    v = new T[newlength];
    assert(v);
```

```
    lngth = newlength;
  }
}

template <class T> inline
void vector<T>::swap(const unsigned n1,const unsigned n2)
{
  assert((n1<lngth) || (n2<lngth));
  T dummy = v[n1]; v[n1] = v[n2]; v[n2] = dummy;
}

template <class T>
void vector<T>::fill(const T& x)          // set all elements = 0
{
  unsigned i;
  for(i=0; i<lngth; i++)
    v[i] = x;
}

template <class T> inline
void vector<T>::unit(const unsigned i)   // sets v to unit vector e_i
{ assert(i<lngth); fill(0); v[i]=1; }

template <class T> inline
unsigned vector<T>::length() const  // return length of vector
{ return lngth; }

template <class T> T vector<T>::maxel() const
{
  T max = 0.0;
  unsigned i;
  for(i=0; i<lngth; i++)
    if(max < v[i]) max = v[i];
  return max;
};

template <class T> T vector<T>::maxabsel() const
{
  T rv2, max = 0.0;
  unsigned i;
  for(i=0; i<lngth; i++)
  {
    rv2 = sqrt(v[i]*v[i]);
    if(max < rv2) max = rv2;
  }
```

```
  return max;
};

template <class T> T vector<T>::norm() const
{
  T sum = 0.0;
  unsigned i;
  for(i=0; i<lngth; ++i)
    sum += v[i]*v[i];
  return sqrt(sum);
}

template <class T> void vector<T>::normalize()
{ *this /= norm(); }

//              Assignment Operator
template <class T>
vector<T>& vector<T>::operator=(const vector<T>& v1)
{
  if(this != &v1)  // check for self assignment (dangerous
  {                // because of delete []v) in resize)
    resize(v1.length());
    unsigned i;
    for(i=0; i<lngth; i++)
      v[i] = v1[i];
  }
  return *this;
}

//                  Operators for class vector
template <class T> inline
T& vector<T>::operator [] (const unsigned pos) const
{ assert(pos<lngth); return v[pos]; }

template <class T>
vector<T> vector<T>::operator + (const vector<T>& v1) const
{
  vector<T> v2(v1);
  assert(v1.length() == lngth);
  unsigned i;
  for(i=0;i<lngth;i++)
    v2[i] += v[i];
  return v2;
}
```

```
template <class T>
vector<T> vector<T>::operator - (const vector<T>& v1) const
{
  vector<T> v2(lngth);
  assert(v1.length() == lngth);
  unsigned i;
  for(i=0;i<=lngth-1;i++)
    v2[i] = v[i] - v1[i];
  return v2;
}

template <class T>
T vector<T>::operator % (const vector<T>& v1) const
{
  T result = 0.0;
  assert(v1.length() == lngth);
  unsigned i;
  for(i=0; i<=lngth-1; ++i)
    result += v1[i] * v[i];
  return result;
}

template <class T>
vector<T> vector<T>::operator * (const vector<T>& v1) const
{
  vector<T> v2(v1);
  assert(v1.length() == lngth);
  unsigned i;
  for(i=0;i<lngth;i++)
    v2[i] *= v[i];
  return v2;
}

template <class T>
vector<T> vector<T>::operator + (const T& x) const
{
  vector<T> v1(lngth);
  unsigned i;
  for(i=0; i<=lngth-1; i++)
    v1[i] = v[i] + x;
  return v1;
}

template <class T>
vector<T> vector<T>::operator - (const T& x) const
```

```
{
  vector<T> v1(lngth);
  unsigned i;
  for(i=0; i<=lngth-1; i++)
    v1[i] = v[i] - x;
  return v1;
}

template <class T>
vector<T> vector<T>::operator * (const T& x) const
{
  vector<T> v1(lngth);
  unsigned i;
  for(i=0; i<=lngth-1; i++)
    v1[i] = v[i]*x;
  return v1;
}

template <class T>
vector<T> vector<T>::operator / (const T& x) const
{
  vector<T> v1(lngth);
  unsigned i;
  for(i=0; i<=lngth-1; i++)
    v1[i] = v[i]/x;
  return v1;
}

template <class T> inline
void vector<T>::operator += (const vector<T>& v1)
{ *this = *this+v1; }

template <class T> inline
void vector<T>::operator -= (const vector<T>& v1)
{ *this = *this-v1; }

template <class T> inline
void vector<T>::operator *= (const vector<T>& v1)
{ *this = *this*v1; }

template <class T> inline
void vector<T>::operator += (const T& x)
{ *this = *this+x; }

template <class T> inline
```

```
void vector<T>::operator -= (const T& x)
{ *this = *this-x; }

template <class T> inline
void vector<T>::operator *= (const T& x)
{ *this = *this*x; }

template <class T> inline
void vector<T>::operator /= (const T& x)
{ *this = *this/x; }

template<class T> vector<T>& vector<T>::operator - ()
{
  unsigned i;
  for(i=0; i<lngth; ++i)
  v[i] = -v[i];
  return *this;
}

template<class T> inline vector<T>& vector<T>::operator + ()
{  return *this; }

//                Output stream access via <<
template<class T>
ostream& operator<<(ostream& os,const vector<T>& vec)
{
  os << "( ";
  unsigned n;
  for(n=0; n<vec.lngth; n++) // Since friend, <<
    os << vec.v[n] << ' ';  // has access to private fields
  return os << ")";
}

//                 Globally defined operators
template<class T>  // for x*vec;
vector<T> operator * (const T& x,const vector<T>& vec)
{ return vec*x; }

void main()
{
  vector<double> v1(4); v1[0]=1.1;v1[1]=2.2;v1[2]=3.3;v1[3]=4.4;
  vector<double> v2(v1);    // using copy constructor
  vector<double> v3=v1+v2;  // using default constructor and =

  cout << v1 << " + " << v2 << " = " << v3 << endl;
```

```
  cout << v1 << " - " << v2 << " = " << v1-v2 << endl;
  cout << v1 << " * " << v2 << " = " << v1*v2 << endl;
  cout << v1 << " % " << v2 << " = " << v1%v2 << endl;
  cout << "-v1 = " << -v1 << endl;   // unary minus operator

  double t = 3.0;
  cout << "v1+t = " << v1+t << endl;
  cout << "v1*t = " << v1*t << endl; // calls the member operator
  cout << "t*v1 = " << t*v1 << endl; // calls the global operator
                                     // which calls v1*t
  v2 += v1;
  v1.normalize();
  cout << "normalized vector v1 = " << v1 << endl;
  cout << "dot product: " << v1%v1 << endl;

  vector<double> v4(2);
  v4 = v1;  v4 = v4;

  cout << "v4 = " << v4 << endl;
  cout << "v1[4] = " << v1[4] << endl;  // out of bounds operation
}
```

The output of this program is given below

```
( 1.1 2.2 3.3 4.4 ) + ( 1.1 2.2 3.3 4.4 ) = ( 2.2 4.4 6.6 8.8 )
( 1.1 2.2 3.3 4.4 ) - ( 1.1 2.2 3.3 4.4 ) = ( 0 0 0 0 )
( 1.1 2.2 3.3 4.4 ) * ( 1.1 2.2 3.3 4.4 ) = ( 1.21 4.84 10.89 19.36 )
( 1.1 2.2 3.3 4.4 ) % ( 1.1 2.2 3.3 4.4 ) = 36.3
-v1 = ( -1.1 -2.2 -3.3 -4.4 )
v1+t = ( 1.9 0.8 -0.3 -1.4 )
v1*t = ( -3.3 -6.6 -9.9 -13.2 )
t*v1 = ( -3.3 -6.6 -9.9 -13.2 )
normalized vector v1 = ( -0.182574 -0.365148 -0.547723 -0.730297 )
dot product: 1
v4 = ( -0.182574 -0.365148 -0.547723 -0.730297 )

Assertion failed:  pos<lngth,  file vector.cpp,  line 198
Abnormal program termination
```

8.4 Matrix Class

8.4.1 Introduction

Matrices are two-dimensional arrays with a certain number of rows and columns. Vectors are included as special case. In this case the number of rows or columns is equal to one. Instead of defining a matrix class from scratch we build it upon our vector class by defining the matrix as a vector of vectors.

```
template <class T> class matrix
{
  protected:
    vector<T>* m;
    unsigned nrows, ncols;
   ...
}
```

Here `m` is a pointer to a vector of type `T`. Defining the matrix as a vector of vectors has the advantage that the matrix class methods can use many of the vector operations defined for the vector class. For example, for matrix addition we simply use vector addition for each of the rows of the matrix. In this way we reduce the amount of code duplication. The following matrix operations are implemented: addition, subtraction, multiplication of two matrices, multiplication of a matrix with a vector or a scalar. We also allow the matrix to be divided by a scalar and we define the unary `+` and `-` operators enabling the user to write statements of the form `B = -A;`, where `A` and `B` are matrices. Furthermore we implement the trace (i.e. the sum of the diagonal elements of a square matrix) and transpose of a matrix. We also define a member function `unit()` which provides a unit matrix and a member function `fill()` which allows to set all entries of the matrix equal to a certain value. We overload the subscription operator such that `M[i]` returns the i'th row as a vector. Adding a second subscription operator `M[i][j]` will call the subscription operator of the vector object for row `i`, returning the `j`'th element of the `i`'th row vector.

Consider the following constructor which declares a matrix of `nrws` rows and `ncls` columns:

```
template<class T>
matrix<T>::matrix(const unsigned nrws,const unsigned ncls)
{
   nrows = nrws;  ncols = ncls;
   m = new vector<T>[nrows];
   assert(m);
   unsigned nr;
   for(nr=0; nr<nrows; nr++)
   m[nr].resize(ncols);
}
```

The statement `m = new vector<T>[nrows];` declares an array of `nrows` elements of objects of the class `vector<T>`, and then sets the pointer variable `m` to the start of the memory area of that array. Implicitly the default constructor is called for each vector object – that is why we have to declare a default constructor for the class if we want to enable the user to declare arrays of objects of the class. However, each vector is now initialized to one of zero length (which is what the default constructor for the vector class does). Thus the public member function `resize()` has to be used to resize each of the row vectors of the matrix. Many of the member functions and operators of the matrix class make use of vector operations. For example, the method `unit()`, which sets the matrix equal to a unit matrix makes uses of the vector method `unit(i)` which sets the vector equal to the unit vector $\mathbf{e}_i$:

```
template<class T>
void matrix<T>::unit()
{
  unsigned nr;
  assert(nrows == ncols);
  for(nr=0; nr<nrows; nr++)
    m[nr].unit(nr);
}
```

The unit vectors are given by $\mathbf{e}_i := (0, ..., 0, 1, 0, ..., 0)$, where 1 stands at the i-th position. Similarly, for matrix addition we use vector addition for each of the row vectors of the matrix:

```
template<class T>
matrix<T> matrix<T>::operator+(const matrix<T>& m1) const
{
  assert((ncols == m1.cols()) || (nrows == m1.rows()));
  matrix<T> m2(m1);
  unsigned nr;
  for(nr=0; nr<nrows; nr++)
    m2[nr] += m[nr];       // make use of vector addition
  return m2;
}
```

The macro `assert` can be used to test the validity of an assertion in a C++ program. If the symbol `NDEBUG` is defined `#define NDEBUG`, then the macro is ignored and no code is generated. To use this macro, the file `assert.h` which contain the macro definition will need to be included.

The subscription is defined such that `m[i]` returns a reference to the `i`'th row vector of the matrix.

```
template<class T> inline
vector<T>& matrix<T>::operator[](const unsigned nrow) const
{ assert (nrow<nrows); return m[nrow]; }
```

This enables the user to make the following statements:

```
matrix<int> m(6,5);
vector<int> v(5);
v = m[2];
m[3] = v;
```

For convenience we define the operator () (which is usually used for function calling) such that m(j) returns the j'th column as a vector.

```
vector<T> matrix<T>::operator()(const unsigned ncol) const
{    // returns column ncol as vector; m[i] is fast, but m(i) is not
  vector<T> colvec(nrows);
  unsigned nr;
  for(nr=0; nr<nrows; nr++)
    colvec[nr] = m[nr][ncol];
  return colvec;
}
```

This operator is of course much slower than the subscription operator since it has to copy the relevant matrix elements into the vector and then return the local vector object by value.

The assignment operator first releases the memory of the matrix, then declares an array of nrows vectors for the rows of the matrix and assigns each of these vectors to the row vectors of the right hand side matrix (the resizing of the vectors is done by the assignment operator defined for the vector class):

```
template<class T>
matrix<T>& matrix<T>::operator=(const matrix<T>& m1)
{
  if(this != &m1)  // check for self assignment
  {                // (dangerous because of delete []m)
    if(m) delete []m;
    nrows = m1.rows();    ncols = m1.cols();
    m = new vector<T>[nrows];
    assert(m);
    unsigned nr;
    for(nr=0; nr<nrows; nr++)
    m[nr] = m1[nr];
  }
  return *this;
}
```

8.4.2 Listing of Matrix Class

```
// MATRIX.CPP
// matrix<T> A(nr,nc); defines a matrix of a built-in
// or user-defined data type, T, with +, -, * for
// matrix multiplication etc., transpose(), trace() and vec*mat.
// The matrix elements are accessed via A[0][0],...,A[nr-1][nc-1].
// A[i] returns row i-1 as vector and A(i) returns column i-1 as vector.
//
// Use #define NDEBUG to switch of bound checking and other
// checking options.

#include <mem.h>
#include <math.h>
#include <assert.h>
#include <iostream.h>
#include "vector.cpp"

template<class T> class matrix
{
  protected:
    vector<T>* m;
    unsigned nrows, ncols;
  public:
     matrix();                    // Default Constructor
     matrix(const matrix& m1); // Copy Constructor
     matrix(const unsigned nrws, const unsigned ncls);
    ~matrix();                    // Destructor

    // Constant Member Functions
    unsigned rows() const;    // returns number of rows
    unsigned cols() const;    // returns number of columns
    T trace() const;          // calculate trace

    // Non-Constant Member Functions
    void unit();              // make matrix a unit matrix
    void fill(const T& x);  // make all elements = x
    matrix<T> transpose();  // returns transpose of m
    void swaprows(const unsigned nr1,const unsigned nr2);
    // swap rows nr1 and nr2

    vector<T>& operator[](const unsigned nrow) const;
      // returns row nrow as vector; used also to access m[i][j]
```

```
    vector<T> operator()(const unsigned ncol) const ; // returns
      // column ncol as vector; NOTE: m[i] is fast, m(i) is not.

    matrix<T>& operator = (const matrix<T>& m1);
    matrix<T>  operator + (const matrix<T>& m1) const;
    matrix<T>  operator - (const matrix<T>& m1) const;
    matrix<T>  operator * (const matrix<T>& m1) const;
    vector<T>  operator * (const vector<T>& v) const;
    matrix<T>  operator + (const T& x) const; // M + x*I
    matrix<T>  operator - (const T& x) const; // M - x*I
    matrix<T>  operator * (const T& x) const; // M * x*I
    matrix<T>  operator / (const T& x) const; // M / x*I
    void       operator += (const matrix<T>& m1);
    void       operator -= (const matrix<T>& m1);
    void       operator *= (const matrix<T>& m1);
    void       operator += (const T& x); // M += x*I
    void       operator -= (const T& x); // M -= x*I
    void       operator *= (const T& x); // M *= x*I
    void       operator /= (const T& x); // M /= x*I

    // Friends of the class
    friend ostream&  operator << (ostream& s,const matrix<T>& m1);
    friend vector<T> operator * (const vector<T>& v,
                                 const matrix<T>& m1);
    friend matrix<T>& operator + (matrix<T>& m1); // unary +
    friend matrix<T>& operator - (matrix<T>& m1); // unary -
};

//               Implementation of Constructors
template <class T> inline
matrix<T>::matrix(): nrows(0), ncols(0), m(NULL) {}

template<class T>
matrix<T>::matrix(const unsigned nrws,const unsigned ncls)
{
  nrows = nrws;  ncols = ncls;
  m = new vector<T>[nrows];
  assert(m);
  unsigned nr;
  for(nr=0; nr<nrows; nr++)
    m[nr].resize(ncols);
}

template<class T> inline
matrix<T>::matrix(const matrix<T>& m1)
```

```
{ m=NULL; *this = m1; }  // see assignment operator

//                Implementation of Destructor
template <class T> inline matrix<T>::~matrix()
{ if (m) delete []m; }

//                 Constant Member Functions
template <class T> inline unsigned matrix<T>::rows() const
{ return nrows; }

template <class T> inline unsigned matrix<T>::cols() const
{ return ncols; }

template<class T> T matrix<T>::trace() const
{
  T sum = 0;
  unsigned nr;
  for(nr=0; nr<nrows; nr++)
    sum += m[nr][nr];
  return sum;
}

//             Non-Constant Member Functions
template<class T>
void matrix<T>::unit()
{
  assert(nrows == ncols);
  unsigned nr;
  for(nr = 0; nr<nrows; nr++)
    m[nr].unit(nr);
}

template<class T>
void matrix<T>::fill(const T& x)
{
  unsigned nr;
  for(nr=0; nr<nrows; nr++)
    m[nr].fill(x);
}

template<class T>
matrix<T> matrix<T>::transpose()
{
  matrix<T> mtransp(ncols,nrows);
  unsigned nr;
```

```
  for(nr=0; nr<nrows; nr++)
  {
    unsigned nc;
    for(nc=0; nc<ncols; nc++)
      mtransp[nc][nr] = m[nr][nc];
  }
  return mtransp;
}

template <class T> inline
void matrix<T>::swaprows(const unsigned nr1,const unsigned nr2)
{
  assert((nr1<nrows) || (nr2<nrows));
  vector<T> dummy = m[nr1];
  m[nr1] = m[nr2];
  m[nr2] = dummy;
}

//      Operators for Accessing rows, columns, elements
template<class T> inline
vector<T>& matrix<T>::operator[](const unsigned nrow) const
{ assert (nrow<nrows); return m[nrow]; }

template<class T>
vector<T> matrix<T>::operator()(const unsigned ncol) const
{    // returns column ncol as vector; NOTE: m[i] is fast,
  vector<T> colvec(nrows);              //  but  m(i) is not.
  unsigned nr;
  for(nr=0; nr<nrows; nr++)
    colvec[nr] = m[nr][ncol];
  return colvec;
}

//                Assignment operator
template<class T>
matrix<T>& matrix<T>::operator=(const matrix<T>& m1)
{
  if(this != &m1)  // check for self assignment
  {                  // (dangerous because of delete []m)
    if(m) delete []m;
    nrows = m1.rows();
    ncols = m1.cols();
    m = new vector<T>[nrows];
    assert(m);
    unsigned nr;
```

```
    for(nr=0; nr<nrows; nr++)
      m[nr] = m1[nr];
  }
  return *this;
}

//                  Arithmetic Operators
template<class T>
matrix<T> matrix<T>::operator+(const matrix<T>& m1) const
{
  assert((ncols == m1.cols()) || (nrows == m1.rows()));
  matrix<T> m2(m1);
  unsigned nr;
  for(nr=0; nr<nrows; nr++)
    m2[nr] += m[nr];       // make use of vector addition
  return m2;
}

template<class T>
matrix<T> matrix<T>::operator-(const matrix<T>& m1) const
{
  assert((ncols == m1.cols()) || (nrows == m1.rows()));
  matrix<T> m2(nrows,ncols);
  unsigned nr;
  for(nr=0; nr<nrows; nr++)
    m2[nr] = m[nr] - m1[nr]; // make use of vector subtraction
  return m2;
}

template<class T>
matrix<T> matrix<T>::operator*(const matrix<T>& m1) const
{
  unsigned ncols1 = m1.cols();
  assert(ncols == m1.rows());
  T sum;
  matrix<T> m2(nrows,ncols1);
  unsigned nr;
  for(nr=0; nr<nrows; nr++)
  {
    unsigned nc1;
    for(nc1=0; nc1<ncols1; nc1++)
    {
      sum = 0;
      unsigned n;
      for(n=0; n<ncols; n++)
```

```
      sum += m[nr][n]*m1[n][nc1];
      m2[nr][nc1] = sum;
    }
  }
  return m2;
}

template<class T>
vector<T> matrix<T>::operator*(const vector<T>& v) const
{
  assert(ncols == v.length());
  vector<T> v1(nrows);
  unsigned nr;
  for(nr=0; nr<nrows; nr++)
  v1[nr] = m[nr]%v;
  return v1;
}

template<class T> matrix<T> matrix<T>::operator+(const T& x) const
{
  assert(nrows == ncols);
  matrix<T> m2(*this);
  unsigned nr;
  for(nr=0; nr<nrows; nr++)
  m2[nr][nr] += x;
  return m2;
}

template<class T> matrix<T> matrix<T>::operator-(const T& x) const
{
  assert(nrows == ncols);
  matrix<T> m2(*this);
  unsigned nr;
  for(nr=0; nr<nrows; nr++)
  m2[nr][nr] -= x;
  return m2;
}

template<class T> matrix<T> matrix<T>::operator*(const T& x) const
{
  matrix<T> m2(nrows,ncols);
  unsigned nr;
  for(nr=0; nr<nrows; nr++)
    m2[nr] = m[nr]*x;     // use vector multiplication
  return m2;
```

```
}

template<class T> matrix<T> matrix<T>::operator/(const T& x) const
{
  matrix<T> m2(nrows,ncols);
  unsigned nr;
  for(nr=0; nr<nrows; nr++)
    m2[nr] = m[nr]/x;     // use vector division
  return m2;
}

template<class T> inline
void matrix<T>::operator += (const matrix<T>& m1)
{ *this = *this+m1; }

template<class T> inline
void matrix<T>::operator -= (const matrix<T>& m1)
{ *this = *this-m1; }

template<class T> inline
void matrix<T>::operator *= (const matrix<T>& m1)
{ *this = *this*m1; }

template<class T> inline
void matrix<T>::operator += (const T& x)   // M + x*I
{ *this = *this+x; }

template<class T> inline
void matrix<T>::operator -= (const T& x)   // M - x*I
{ *this = *this-x; }

template<class T> inline
void matrix<T>::operator *= (const T& x)   // M *= x*I
{ *this = *this*x; }

template<class T> inline
void matrix<T>::operator /= (const T& x)   // M /= x*I
{ *this = *this/x; }

//                  Friends of class matrix
template<class T>
ostream& operator<<(ostream& s,const matrix<T>& m)
{
  unsigned nr;
  for(nr=0; nr<m.rows(); ++nr)
```

```
  {
    unsigned nc;
    s << "( ";
    for(nc=0; nc<m.cols(); nc++)
    {
    s.precision(5);
    s.width(8);
    s << m[nr][nc] << ' ';
    }
    s << ")" << endl;
  }
  return s;
}

template <class T>
vector<T> operator * (const vector<T>& v,const matrix<T>& m1)
{
  int ncls = m1.ncols;  int nrws = m1.nrows;
  assert(nrws == v.length());
  vector<T> v1(ncls);

  unsigned nc;
  for(nc=0; nc<ncls; nc++)
  {
  v1[nc]= 0.0;
  unsigned nr;
  for(nr=0; nr<nrws; nr++)
  v1[nc] += v[nr]*m1[nr][nc];
  }
  return v1;
}

template<class T> matrix<T>& operator - (matrix<T>& m1)
{
  unsigned i;
  for(i=0; i<=m1.rows()-1; ++i)
  m1[i] = -m1[i];  // using unary minus for vectors
  return m1;
}

template<class T> inline vector<T>& operator + (matrix<T>& m1)
{ return m1; }

//             main function
void main()
```

```
{
  unsigned nd;
  cout << "Enter dimension of matrices: ";
  cin  >> nd;

  matrix<double> A(nd,nd), B(nd,nd), C(nd,nd);
  vector<double> v(nd);
  double determ;

  int i;
  for(i=0; i<nd; ++i)
  {
    int j;
    for(j=0; j<nd; ++j)
    {
    A[i][j] = 2*i+j;
    B[i][j] = i-j;
    }
    v[i] = 2*i+3;
  }
  C = A*B;
  cout << "A = " << endl << A;
  cout << "B = " << endl << B;
  cout << "A+B = " << endl << A+B;
  cout << "A*B (dot product) = " << endl << C;
  cout << "A*v = " << A*v << "   but   v*A = " << v*A << endl;

  A *= B;
  cout << "A *= B => " << endl << A;
  cout << "A.transpose() = " << endl << A.transpose();

  cout << "trace(A) = " << A.trace();
}
```

8.5 Linked Lists

8.5.1 Introduction

Many of the abstract data types can be viewed as special cases of a more general abstract data type, the *linked list.* A stack, for example, is readily described and implemented as a linked list, as are strings and sets. We study the linked list abstract data type and see how it can be applied to these and other situations. Linked lists are useful whenever the number of elements of a collection varies regularly. For example, a bank constantly closes certain accounts and opens other accounts. One thus needs a data structure which allows one to add and remove elements conveniently. A singly linked list has a pointer to the first node and each node has its own data fields and methods and a pointer to the next node. The pointer of the final node is set to NULL. A definition of a linked list is:

Definition. A linked list is a sequence of zero or more nodes, together with a set of operations. Each node contains two fields, an item field that holds information and a next field that holds a pointer to the next node of the list.

We consider singly linked list, circular linked lists, doubly linked lists, recursive linked lists, intrusive linked lists and generalized lists. Increased flexibility and processing speed is usually obtained at the cost of higher memory requirement. For example, a double linked list allows easy forward and backward traversing of the list. In a single linked list going one element back requires that we start at the first element and step through the list until the next pointer points to the current element. The price we have to pay for this faster access is that we have to store for each element two pointers, one pointing to the next node and one pointing to the previous node. The generalized linked list (i.e. a list which also allows lists as elements) is the basic abstract data type of some languages. The best known example is the language LISP, which stands for LIST PROCESSING.

The following two programs show the basic idea how a linked list is built.

```
// list1.cpp

#include <iostream.h>

class List
{
  public:
     int data;
     List *next;
};

int main()
{
    List a, b, c;  // a, b, c objects of class List
    // Assignment
    a.data = 7;  b.data = 3;  c.data = 9;
    a.next = b.next = c.next = NULL;

    // We now chain the three classes together
    a.next = &b;  b.next = &c;
    // These pointer assignments result in linking a to b to c

    // The links allow us to retrieve data from successive elements
    cout << a.next -> data << endl;         // output 3
    cout << a.next -> next -> data << endl; // output 9
    cout << b.next -> data << endl;         // output 9
    int i = a.next -> data;
    cout << i << endl;                      // output 3
    cout << a.data << endl;                 // output 7
    return 0;
}
```

Each variable of the class `List` has two members, `data` and `next`. The pointer variable `next` is called a *link*. Each structure is linked to a succeeding structure by way of the member `next`. The pointer variable `next` contains an address of either the location in memory of the successor `List` element or the special value `NULL`, which is defined in `<iostream.h>` as a symbolic constant with the value 0. The value `NULL` is used to denote the end of the list.

```
// list2.cpp

#include <iostream.h>
#include <string.h>

class Person
{
   public:
      char name[10];
      Person* next;
};

void print(Person* ptr)
{
   int count = 1;
   cout << endl << endl;
   while(ptr != NULL)
   {
      cout << "Name number is: " << count++ << " " << ptr -> name;
      cout << endl;
      ptr = ptr -> next;
   }
}

int main()
{
   Person name1, name2, name3, *start;

   strcpy(name1.name,"Otto");
   strcpy(name2.name,"Ludwig");
   strcpy(name3.name,"Hans");

   name1.next = &name2;  // Otto points to Ludwig
   name2.next = &name3;  // Ludwig points to Hans
   name3.next = NULL;    // Hans is last

   start = &name1;
   print(start);

   cout << endl;

   return 0;
}
```

When the function print is invoked

```
print(start);
```

the value of start is copied into the parameter ptr. Then the variable count is initialized to 1:

```
int count = 1;
```

At the while loop

```
while(ptr != NULL) {
```

we first check whether the value of ptr is NULL. C and C++ guarantees that NULL is not an actual address. When the value of ptr is equal to NULL, we know that ptr cannot point to another node and that we are, in fact, at the end of the linked list. In our case, ptr is not equal to NULL. Thus we execute the body of the while loop. The statement

```
cout << "Name number is: " << count++ << " " << ptr -> name;
```

prints `Name number is: 1 Otto` and increments the value of count. Next, we execute

```
ptr = ptr -> next;
```

The value of `ptr -> next` is the address of name2. Thus we copy the address of name2 into ptr. The effect is to cause ptr to point to name2. An expression such as `ptr = ptr -> next;` occurs frequently when processing linked lists. The effect is move the pointer ptr to the next node in the linked list. We return to the top of the while loop. Again the expression

```
ptr != NULL
```

is true, so we execute the body of the while loop. The statement

```
cout << "Name number is: " << count++ << " " << ptr -> name;
```

prints `Name number is: 2 Ludwig` and increments the value of count by one. Next, we execute `ptr = ptr -> next;` which points ptr to name3. We return to the top of the while loop. Again the expression `ptr != NULL` is true, so we execute the body of the while loop. The first statement

```
cout << "Name number is: " << count++ << " " << ptr -> name;
```

prints `Name number is: 3 Hans` and increments the value of count by one. Next, we execute

```
ptr = ptr -> next;
```

Since the value of `ptr -> next` is NULL, we assign to the pointer ptr the value NULL. We return to the top of the while loop. Since the value of `ptr -> next` is NULL, we assign to ptr the value NULL. We return to the top of the while loop. Since `ptr != NULL` is false, we terminate the loop, and the function returns. The program terminates.

8.5.2 Simple Linked Lists

Here we give a collection of programs of simple linked lists. In program `list1.cpp` the member function function `display` display the list on the screen. The member function `additem` adds an item (here an integer) to the list.

```
// list1.cpp
#include <iostream.h>

class Link
{
   public:
     int data;      // data item (integer)
     Link* next;    // pointer to next link
};

class List      // a list of links
{
   private:
     Link* first;   // pointer to first link
   public:
     List();        // constructor with no argument
     void additem(int);  // add data item (one link)
     void display();       // display all links
};

List::List()
{
   first = NULL;             // no first link
   cout << "List constructor invoked " << endl;
}

void List::additem(int d) // add data item
{
   Link* newlink;                 // declare pointer to Link
   newlink = new Link;            // make a new Link
   newlink -> data = d;           // give it data
   newlink -> next = first;       // it points to next link
   first = newlink;               // now it points to this
}

void List::display()            // display of links
{
   Link* current;               // pointer to Link
   current = first;             // set pointer to first link
```

```
   while(current != NULL)      // quit on last link
   {
   cout << endl << current -> data; // display data
   current = current -> next;       // move to next Link
   }
}

int main()
{
   List li;      // create linked list

   li.additem(14);
   li.additem(90);
   li.additem(45);
   li.additem(5);

   li.display();

   return 0;
}
```

The output is

```
5  45  90  14
```

8.5.3 Singly Linked Circular List

It is quite easy to join the end of a linear list back to its head. One has simply to replace the NULL in the last node with a pointer back to the first node. Then we have a *singly linked circular list.*

As an application consider the following problem. A deranged king has decided to choose his successor in most brutal way. He directs his m knights to sit at his round table. Then he numbers them $1, 2, 3, ..., m$ clockwise. The king then chooses a cycle number, k, and starting with the first knight, counts off $1, 2, 3, ..., k$. The k-th knight is carried off and summarily executed, and count-off by k continues with the next surviving knight. The process goes on until there is but one survivor. He is the heir to the throne. For example, let $m = 7$ (the number of knights) and $k = 3$. Then the survivor would be knight number 4. This problem can be formulated as a singly linked circular list.

```
// king.cpp

#include <iostream.h>

class Knight
{
   public:
      int number;
      Knight* next;
};

int S(int m, int k)
{
   Knight* first;
   first = NULL;
   Knight* current;
   current = NULL;
   Knight* ptr;
   ptr = NULL;

   int i;
   for(i=0; i < m; i++)
   {
      ptr = new Knight;
      if(first == NULL)
      {
      first = ptr;
      current = ptr;
```

```
      current -> number = i + 1;
      }
      else {
      current -> next = ptr;
      current = ptr;
      current -> number = i + 1;
      }
    }

    current -> next = first;
    int cnt;
    current = first;
    Knight* prevKnight;
    prevKnight = NULL;

    while(current -> next != current)
    {
      for(cnt = 0; cnt < k-1; cnt++)
      {
      prevKnight = current;
      current = current -> next;
      }
      prevKnight -> next = current -> next;
      delete current;
      current = prevKnight -> next;
      }
   return current -> number;
}

int main()
{
   int m, k, t;
   cout << endl << "enter number of knights: ";
   cin >> m;
   cout << endl << "enter cycle number: ";
   cin >> k;
   cout << endl << endl;
   cout << "the survivor is knight number: " << S(m,k);
   cout << endl;

   return 0;
}
```

8.5.4 Recursive List

The program `list2.cpp` implements the list as a recursive list. Thus this implementation does not include the class Node. The member function `reverse` reverses the list recursively.

```
// list2.cpp

#include <iostream.h>
#include <stdlib.h>
#include <assert.h>  // for assert

typedef int T;

 // Declaration class List
class List
{
   public:
      List();
      ~List();

      void Insert(const T&);
      int Search(const T&);
      int Delete(const T&);

      T Head();
      List* Tail();
      int Empty();
      List* Reverse(List*);
    private:
      T head;
      List* tail;
      int empty;
}; // List

List::List()
{
   empty = 1;
}

List::~List()
{
   if(!empty)
     delete tail;
}
```

```
void List::Insert(const T &toInsert)
{
   if(empty)
   {
      head = toInsert;
      tail = new List;
      empty = 0;
   }
   else
      tail->Insert(toInsert);
}

int List::Search(const T &toSearch)
{
   if(empty)  return 0;
   else if(head == toSearch)  return 1;
   else return tail -> Search(toSearch);
}

T List::Head()
{
   assert(!empty);
   return head;
}

List* List::Tail()
{
   assert(!empty);
   return tail;
}

int List::Empty()
{
   return empty;
}

List* List::Reverse(List* L)
{
   if(L -> Empty())
   {
     List* temp;
     temp = new List;
     return temp;
   }
```

```
   else
   {
      List* R;
      R = Reverse(L->Tail());
      (*R).Insert(L->Head());
      return R;
   }
}

int main()
{
   List L;
   int i;
   for(i=1; i<=8; i++)  L.Insert(i);

   List* LX = &L;
   cout << "The initial List is: " << endl;
   while(!LX -> Empty())
   {
      cout << LX -> Head() << ' ';
      LX = LX -> Tail();
   }
   cout << endl << endl;

   List* R = L.Reverse(&L);
   List* LP = R;

   while(!LP -> Empty())
   {
      cout << LP->Head() << ' ';
      LP = LP -> Tail();
   }
   cout << endl << endl;

   cout << "what happened to the initial list: " << endl;
   LP = &L;
   while(!LP -> Empty())
   {
      cout << LP -> Head() << ' ';
      LP = LP -> Tail();
   }
   cout << endl;
   return 0;
}
```

8.5.5 Linked List and Templates

In this section we give a template class for a link list.

```
// list3.cpp

#include <iostream.h>

template<class T> class Node {
   private:
     T Value;
   public:
     Node();
     Node(T);
     void display();
     Node<T> *next;
};

template<class T> Node<T>::Node()  { }

template<class T> Node<T>::Node(T value) {  Value = value; }

template<class T> void Node<T>::display() { cout << Value << " "; }

template<class T> class List {
   private:
     Node<T> first;
     Node<T> *last;
   public:
     List();
     void show_list();
     void append(Node<T> *new_node);
};

template<class T> List<T>::List()
{
   first.next = NULL;
   last = &first;
}

template<class T> void List<T>::show_list()
{
   Node<T>* current;
   current = first.next;
   while(current)
```

```
   {
   current -> display();
   current = current -> next;
   }
}

template<class T> void List<T>::append(Node<T> *new_node)
{
  last -> next = new_node;
  last = new_node;
  last -> next = NULL;
}

int main()
{
   List<int> single;
   List<float> values;
   List<char> letters;
   List<char*> names;

   int i;
   for(i=1; i <= 10; i++)
   single.append(new Node<int> (i));
   single.show_list();
   cout << endl << endl;

   for(i=1; i <= 10; i++)
   values.append(new Node<float> (0.1*i));
   values.show_list();
   cout << endl << endl;

   for(i = 'A'; i <= 'Z'; i++)
   letters.append(new Node<char> (i));
   letters.show_list();
   cout << endl << endl;

   names.append(new Node<char*> ("willi"));
   names.append(new Node<char*> ("eng"));
   names.append(new Node<char*> ("hans"));
   names.show_list();

   return 0;
}
```

8.5.6 Linked List and Iterators

Typically we want to insert elements, remove elements and push all elements to an output stream (e.g. print all elements):

```
template <class T> class LinkedList        // can be used as is for
{                                          // a stack (LIFO)
  protected:
    Node<T>* _head;
    void newhead(const T& element);

  public:
    LinkedList() : _head(NULL) {};
    LinkedList(const T& element);
    ~LinkedList();
    void push(const T& element);
    T pop();

    friend ostream& operator << (ostream& os,LinkedList<T>& ll);
    friend class LinkedListIterator<T>;
};
```

The only data field of the linked list is a pointer to the first node. Hence all insert and remove operations act on the first node and we have effectively a last-in-first-out (LIFO) structure, which is commonly known as a stack. To make this equivalence obvious, we named the insertion function `push` and the deletion function `pop`. For efficiency reasons we prefer a doubly linked circular list, where each node has a pointer to the next and one to the previous node in the linked list. The data fields of a node are thus these two pointers together with the element which is stored in that node. The latter is declared on a template. Note that we declared `LinkedListIterator` as a friend of the class. This class will enable us to iterate through the list, remove specific elements from the list, let an object stored in a node perform a certain function and so on. The advantages of defining the iterator outside the LinkedList class are the following:

- A linked list can have several iterators, each looking at a different part of the list.
- The iterator `LinkedListIterator` will be defined as a derived class of the base class `Iterator`. Other derived classes would be `IfstreamIterator` and `VectorIterator` and programs can be developed without referring to the base class `Iterator` without being bound to an underlying data structure (i.e. a linked list, a file or a vector).

The nodes have the information about the previous and the following node and they thus define the methods `InsertBefore` and `InsertAfter`. A `Node` can also remove itself from the list by linking its two neighbours directly to one another:

```
template <class T> class Node     // all members private
{
  private:
    Node<T> *_next, *_previous;
    T _element;

    Node();
    void InsertBefore(const T& element);
    void InsertAfter(const T& element);
    void remove();

    friend class LinkedList<T>;
    friend class LinkedListIterator<T>;
    friend ostream& operator << (ostream& os,LinkedList<T>& ll);
};
```

Note that the entire class is declared private including the constructor. Hence, only the friends of this class can use it. The classes which should have access to the nodes are `LinkedList` and `LinkedListIterator`, both of which are declared as friends.

For example, the `LinkedList` method `pop` directly uses the private data fields `_next` and `_element` and the private method `remove()`:

```
template <class T> T LinkedList<T>::pop()
{
  assert(_head != NULL);
  Node<T>* dummy = _head;
  T dumel = dummy->_element;
  _head = _head->_next;

  if(dummy == _head)
  _head = NULL;
  dummy->remove();
  delete dummy;
  return dumel;
};
```

The destructor of the `LinkedList` class repetitively calls `pop()` until the linked list is empty:

```
template <class T> LinkedList<T>::~LinkedList()
{
   while(_head != NULL) pop();
};
```

The complete listing is as follows:

```
// list4.cpp
#include <iostream.h>
#include <stdlib.h>
#include <assert.h>

//            Interface for class LinkedList
template <class T> class LinkedList
{
  protected:
    Node<T>* _head;
    void newhead(const T& element);
  public:
    LinkedList() : _head(NULL) {};
    LinkedList(const T& element);
    ~LinkedList();
    void append(const T& element);
    void push(const T& element);
    T pop();
    friend ostream& operator << (ostream& os,LinkedList<T>& ll);
    friend class LinkedListIterator<T>;
};

//             Interface for class Node
template <class T> class Node     // all members private
{
  private:
    Node<T> *_next, *_previous;
    T _element;
    Node();
    void InsertBefore(const T& element);
    void InsertAfter(const T& element);
    void remove();
    friend class LinkedList<T>;
    friend class LinkedListIterator<T>;
    friend ostream& operator << (ostream& os,LinkedList<T>& ll);
};

//     Interface for class Abstract Base Class Iterator
template <class T> class Iterator
{
  public:
    virtual void reset() = 0;
    virtual void operator ++ () = 0;
    virtual T operator () () const = 0;
    virtual int operator ~ () const = 0;
```

```
    virtual void operator = (const T& x) = 0;
};

//           Interface for class LinkedListIterator
template <class T> class LinkedListIterator : public Iterator<T>
{
  protected:
    const LinkedList<T>& _list;
    Node<T>* _current;
  public:
    LinkedListIterator(LinkedList<T>& list);
    virtual void reset();
    virtual void last();
    virtual void operator ++ ();
    virtual T operator () () const;
    virtual int  operator ~ () const;
    virtual void operator = (const T& x);
    virtual void operator -- ();
    virtual void Insertbefore(const T& x);
    virtual void InsertAfter(const T& x);
    virtual void InsertAfterInc(const T& x);
    virtual void remove();
};

//             Implementation of class LinkedList
template <class T> inline
void LinkedList<T>::newhead(const T& element)
{
  _head = new Node<T>;
  _head->_element = element;
  _head->_next = _head;
  _head->_previous = _head;
}

template <class T> inline LinkedList<T>::LinkedList(const T& element)
{  newhead(element); };

template <class T> LinkedList<T>::~LinkedList()
{  while(_head != NULL) pop(); }

template <class T>
void LinkedList<T>::append(const T& element)
{
  if(_head == NULL)  newhead(element);
  else  _head->InsertBefore(element);
```

```
}

template <class T> void LinkedList<T>::push(const T& element)
{
  append(element);
  _head=_head->_previous;
}

template <class T> T LinkedList<T>::pop()
{
  assert(_head != NULL);
  Node<T>* dummy = _head;
  T dumel = dummy->_element;
  _head = _head->_next;
  if(dummy == _head)
    _head = NULL;
  dummy->remove();
  delete dummy;
  return dumel;
}

template <class T> ostream& operator << (ostream& os,LinkedList<T>& ll)
{
  Node<T>* dummy = ll._head;
  while(dummy->_next != ll._head)
  {
  os << dummy->_element << endl;
  dummy = dummy->_next;
  }
  os << dummy->_element << endl;
  return os;
}

//                  Implementation of class Node
template <class T> inline Node<T>::Node()
{ _next = this; _previous = this; }

template <class T> inline void Node<T>::InsertBefore(const T& element)
{
  Node<T>* newnode = new Node<T>;
  newnode->_element = element;
  newnode->_previous = _previous;
  newnode->_next = this;
  _previous->_next = newnode;
  _previous = newnode;
```

```
}

template <class T> inline void Node<T>::InsertAfter(const T& element)
{
  Node<T>* newnode = new Node<T>;
  newnode->_element = element;
  newnode->_previous = this;
  newnode->_next = _next;
  _next->_previous = newnode;
  _next = newnode;
}

template <class T> inline void Node<T>::remove()
{
  _previous->_next = _next;
  _next->_previous = _previous;
}

//         Implementation of class LinkedListIterator
template <class T> LinkedListIterator<T>::LinkedListIterator
  (LinkedList<T>& list) : _list(list)
{ reset(); }

template <class T> void LinkedListIterator<T>::reset()
{ _current = _list._head; }

template <class T> void LinkedListIterator<T>::last()
{ _current = _list._head->_previous; }

template <class T> void LinkedListIterator<T>::operator ++ ()
{
  if((_current == _list._head->_previous) || (!_current))
    _current = NULL;
  else
    _current = _current->_next;
}

template <class T> T LinkedListIterator<T>::operator () () const
{ return _current->_element; }

template <class T> int LinkedListIterator<T>::operator ~ () const
{ return (_current != NULL); }

template <class T>
void LinkedListIterator<T>::operator =  (const T& x)
```

```
{ _current->_element = x; }

template <class T>
void LinkedListIterator<T>::Insertbefore(const T& x)
{ _current->InsertBefore(x); }

template <class T>
void LinkedListIterator<T>::InsertAfter(const T& x)
{ _current->InsertAfter(x); }

template <class T>
void LinkedListIterator<T>::InsertAfterInc(const T& x)
{
  InsertAfter(x);
  ++*this;
}

template <class T>
void LinkedListIterator<T>::operator -- ()
{
  if((_current == _list._head->_previous) || (!_current))
    _current = NULL;
  else
    _current = _current->_next;
}

template <class T> void LinkedListIterator<T>::remove()
{
  Node<T>* dummy = _current;
  _current = _current->_next;
  dummy->remove();
  delete dummy;
}

//                    class VectorIterator
#include "vector.cpp"
template <class T> class VectorIterator : public Iterator<T>
{
  protected:
    const vector<T>& _vec;
    int _current;
  public:
    VectorIterator(const vector<T>& vec) : _vec(vec), _current(0) {};
    virtual void reset() { _current=0; };
    virtual void operator ++ () { ++_current; };
```

```
    virtual int  operator ~ () const
      { return ((_current >= 0) && (_current < _vec.length())); };
    virtual T operator () () const { return _vec[_current]; };
    virtual void operator = (const T& x) { _vec[_current]=x; };
};

template <class T>
void display(Iterator<T>& iter)
{
  for(iter.reset(); ~iter; ++iter)
    cout << iter() << endl;
}

void main()
{
  LinkedList<int> lst(2);
  LinkedListIterator<int> lstiter(lst);
  int i;
  for(i=1; i<=4; ++i)
  lstiter.InsertAfterInc(i);

  vector<char> vec(2);
  vec[0] = 'A'; vec[1] = 'B';

  VectorIterator<char> veciter(vec);
  display(lstiter);
  cout << endl;
  display(veciter);

  lstiter.reset();
  ++lstiter;
  lstiter.remove();
  cout << endl << lst;
  cout << "popping : " << lst.pop() << endl;
  cout << "pushing : " << 17 << endl; lst.push(17);
  display(lstiter);
}
```

8.5.7 Intrusive List

So far we have looked at non-intrusive linked lists. Each element (node) of a non-intrusive linked list contains a copy of the actual object stored in the linked list. This is perhaps adequate for simple objects (i.e. integers or characters), but if we have huge objects which we want to store in a linked list then this method is highly inefficient. The following listing defines an intrusive doubly-linked list class and at the end of the listing we give a small example program for using this linked list class.

```
// list5.cpp

#include <iostream.h>

class Node
{
  public:
    Node() : next(0), previous(0) {};
    Node(Node* node) : next(node), previous(node) { };
    Node *next, *previous;
};

class BaseLinkList
{
  public:
    BaseLinkList () : head(0) {};
    BaseLinkList (Node* node)
    { head = node -> next = node -> previous = node; };
    void push(Node* node);
    void append(Node* node);
    void clear() { head = 0; };
    Node* pop();
    friend class BaseLinkListIterator;
  protected:
    Node* head;
};

template <class T> class IntrusiveLinkList : public BaseLinkList
{
  public:
    IntrusiveLinkList () : BaseLinkList() {};
    IntrusiveLinkList (T* element) : BaseLinkList(element) {};
    void push(T* element) { BaseLinkList::push(element); };
    void append(T* element) { BaseLinkList::append(element); };
    T* pop (){ return (T*)BaseLinkList::pop(); };
```

```
    friend class IntrusiveLinkListIterator<T>;
 };

void BaseLinkList::append (Node* node)
{
  if(head)
  {
  node -> next = head;
  node -> previous = head -> previous;
  head -> previous -> next = node;
  head -> previous = node;
  }
  else
  {
  head = node;
  head -> next = head;
  head -> previous = head;
  }
};

void BaseLinkList::push(Node* node)
{
  append(node);
  head = head->previous;
};

Node* BaseLinkList::pop()
{
  Node* ReturnNode(0);
  if(head != 0)
  {
    ReturnNode = head;
    if(head->next == head) head = 0;
    else
    {
    head = head -> next;
    head -> next -> previous = head -> previous;
    }
  }
  return ReturnNode;
};

//     Interface for class Abstract Base Class Iterator
template <class T> class Iterator
{
```

```
  public:
    virtual void reset() = 0;
    virtual void operator ++ () = 0;
    virtual T*   operator () () const = 0;
    virtual int  operator ~ () const = 0;
    virtual void operator = (T* x) = 0;
};

//    Interface for class IntrusiveLinkedListIterator
template <class T>
class IntrusiveLinkListIterator : public Iterator<T>
{
  protected:
    IntrusiveLinkList<T>& _list;
    Node* _current;
  public:
    IntrusiveLinkListIterator(IntrusiveLinkList<T>& list) :
    _list(list) {reset();};
    virtual void reset() { _current = _list.head; };
    virtual void last()  { _current = _list.head->previous; };
    virtual void operator ++ ();
    virtual T*   operator () () const;
    virtual int operator ~ () const;
    virtual void operator = (T* x);
    virtual void operator -- ();
    virtual void InsertBefore(T* x);
    virtual void InsertAfter(T* x);
    virtual void InsertAfterInc(T* x);
    virtual void remove();
};

//        Implementation for IntrusiveLinkListIterator
template <class T>
void IntrusiveLinkListIterator<T>::operator ++ ()
{
  if((_current == _list.head->previous) || (!_current))
    _current = 0;
  else
    _current = _current->next;
}

template <class T>
T* IntrusiveLinkListIterator<T>::operator () () const
{ return (T*)_current; }
```

```
template <class T>
int IntrusiveLinkListIterator<T>::operator ~ () const
{ return (_current != 0); }

template <class T>
void IntrusiveLinkListIterator<T>::operator = (T* x)
{ _current = x; }

template <class T>
void IntrusiveLinkListIterator<T>::InsertBefore(T* x)
{
  Node* dummyhead = _list.head;
  _list.head = _current;
  _list.push(x);
  _list.head = dummyhead;
}

template <class T>
void IntrusiveLinkListIterator<T>::InsertAfter(T* x)
{
  Node* dummyhead = _list.head;
  _list.head = _current->next;
  _list.push(x);
  _list.head = dummyhead;
}

template <class T>
void IntrusiveLinkListIterator<T>::InsertAfterInc(T* x)
{
  InsertAfter(x);
  ++*this;
}

template <class T>
void IntrusiveLinkListIterator<T>::operator -- ()
{
  if ((_current == _list.head->previous) || (!_current))
    _current = NULL;
  else
    _current = _current->next;
}

template <class T> void IntrusiveLinkListIterator<T>::remove ()
{
  if(_current == _list.head)
```

```
  {
  if(_current->next == _current)
  _list.head = 0;
  else
  _list.head = _current->next;
  }
  _current->previous->next = _current->next;
  _current->next->previous = _current->previous;
}

class intnode : public Node
{
  public:
    intnode (const int x) : ix(x) { };
    intnode () : ix(0) { };
    int operator () () const { return ix; };
  private:
    int ix;
}

void main()
{
  IntrusiveLinkList<intnode> ilist;
  ilist.append(new intnode(2));
  ilist.append(new intnode(3));
  ilist.append(new intnode(4));
  ilist.append(new intnode(5));
  ilist.append(new intnode(6));

  IntrusiveLinkListIterator<intnode> iter(ilist);
  for(iter.reset(); ~iter; ++iter)
  {
  intnode* in = iter();
  cout << (*in)() << endl;
  }

  iter.reset();
  iter.remove();
  cout << endl;
  for(iter.reset(); ~iter; ++iter)
  {
    intnode* in = iter();
    cout << (*in)() << endl;
  }
}
```

8.5.8 Generalized Lists

Imagine a single linear linked list of four items, written

```
(A  B  C  D)
```

as shorthand for the usual node-and-pointer diagram. Then imagine a list some of whose nodes are themselves lists

```
(A  ((B  C)  D)  E  (F  G))
```

This is a special case of a *generalized list.* It also has four items: the first item is `A`; the second is a sublist made up of the sublist `(B C)` and `D`; the third is `E`; and the fourth is the sublist `(F G)`. The generalized list has proved to be such a useful data structure that it has been made the basis of a very powerful language called LISP.

How might we build a storage structure in C and C++ to hold a linked list with sublists ? At each link we have two pointers to `Node`.

Some of these nodes look quite conventional: the left part has some literal value, `A`, `B`, `C`, and so forth (called an *atom* in LISP), and the right part contains a pointer. However other nodes consists of two pointers. The center field of the node is either a character or a nodepointer (clink is short for center link). The function we need to make generalized lists is `cons`, which constructs a list by adding a new first element to (a copy of) an old one. A list can be started by using `cons` to prefix an atom to `NULL`, the empty list. The function `cons` needs two arguments: the first item to be added to the list, and a pointer to that list. Since the first argument (the new first item) can be either an atom or alist, we do not know what type to put in the function heading. Thus we need two `cons` function, the one is used when the firt node is an atom, and the other one when it is a pointer to a list. A polymorphic generalized list is described and implemented by Tan Kiat Shi and Steeb [14].

In the program we introduce besides the class `Node` a `class List`. The functions `cons`, `car`, `cdr` and `copy` are member functions of `class List`.

```
// lisp.cpp

#include <iostream.h>

class Node
{
   public:
      Node();
      Node(const Node&);
      ~Node();
      int atom;
      char info;
      Node* next;
      Node* clink;
};

class List
{
   private:
      Node* head;
   public:
      List();                          // Constructor
      List(const List&);               // Copy constructor
      ~List();                         // Destructor
      List& operator = (const List&); // overloading assignment
      Node* copy(Node*);
      List cons(char,List&);
      List cons(List&,List&);
      List car(List&);
      List cdr(List&);
      void printNode(Node*);
      void printList();
      void release(Node*);
};

Node::Node() {  }

Node::~Node() {  }

Node::Node(const Node& nd)
{
   atom = nd.atom;
   info = nd.info;
   next = nd.next;
```

```
   clink = nd.clink;
}

List::List() { head = NULL; }

List::List(const List& larg)
{
   if(larg.head == NULL)
       head = NULL;
   else
       head = copy(larg.head);
}

List::~List()
{
   if(head)
   {
   release(head);
   }
   delete head;
}

List& List::operator = (const List& larg)
{
   if(larg.head == NULL)
       head = NULL;
   else
       head = copy(larg.head);
   return *this;
}

Node* List::copy(Node* narg)
{
   Node* res;
   res = new Node;
   if(narg != NULL)
       {
       res->atom = narg->atom;
       if(res->atom)
       {
       res->info = narg->info;
       res->clink = NULL;
       }
       else
       res->clink = copy(narg -> clink);
```

```
    res->next = copy(narg -> next);
    }
    else
    res = NULL;
  return res;
}

List List::cons(char newatom,List& oldlist)
{
  List res(oldlist);
  Node* newNode;
  newNode = new Node;
  newNode -> atom = 1;
  newNode -> info = newatom;
  newNode -> clink = NULL;
  newNode -> next = res.head;
  res.head = newNode;
  return res;
}

List List::cons(List& newList,List& oldList)
{
  List res(oldList);
  Node* newNode;
  newNode = new Node;
  newNode -> atom  = 0;
  newNode -> clink = copy(newList.head);
  newNode -> next  = res.head;
  res.head = newNode;
  return res;
}

List List::car(List& oldList)
{
  List res(oldList);
  List empty;
  List carList;
  if(res.head -> clink == NULL)
  carList=cons(res.head->info,empty);
  else
  carList.head = copy(res.head->clink);
  return carList;
}

List List::cdr(List& oldList)
```

```
{
        List res(oldList);
        res.head = res.head->next;
        return res;
}

void List::printNode(Node* narg)
{
   Node* tmp;
   tmp = copy(narg);
   cout << "(";

   while(tmp != NULL)
       {
       if(tmp->atom)
           cout << tmp->info << " ";
       else
           printNode(tmp->clink);
       tmp = tmp->next;
       }
   cout << ")";
}

void List::printList()
{
   printNode(head);
}

void List::release(Node* ptr)
{
   Node* tmp;
   {
   if((tmp = ptr -> next) != NULL)
   {
   release(tmp);
   delete tmp;
   tmp = NULL;
   }
   if((tmp = ptr -> clink) != NULL)
   {
   release(tmp);
   delete tmp;
   tmp = NULL;
   }
   }
```

```
}

int main()
{
    List empty;
    List L;
    L = L.cons('G', empty);
    cout << "L = ";
    L.printList();
    cout << endl;

    List M;
    M = M.cons('E',M.cons(L,empty));
    cout << "M = ";
    M.printList();
    cout << endl;

    List N;
    N = N.cons(N.cons('B',N.cons('C',empty)),N.cons('D',empty));

    cout << "N = ";
    N.printList();
    cout << endl;

    List P;
    P = P.cons('A',P.cons(N,M));
    cout << "P = ";
    P.printList();
    cout <<endl;

    List Z;
    Z = Z.cons('1',Z.cons(Z.cons('2',Z.cons('7', empty)),
        Z.cons('5',empty)));
    cout << "Z = ";
    Z.printList();
    cout << endl;
    List Zcar = Z.car(Z);
    cout << "car (Z) = ";
    Z.printList();
    cout << endl;
    return 0;
}
```

8.6 Binary Trees

8.6.1 Introduction

The binary search technique makes it possible to search through a large body of data in a very efficient manner. Suppose that there are one million numbers or strings stored in a list and we want to find out whether a particular number or string is not in the list. If the target number or string is not in the list, we have to check all one million numbers or strings to obtain this result. If we apply a binary search technique to search for the numbers or string we can find the number (or find out it is not there) after checking not more than 20 numbers or strings. The trick to binary search is to store data in a constrained manner. The datas are stored in a structure called a *binary tree.*

A definition is as follows. A *binary tree* is a finite set of nodes that is either empty, or consists of a root and two disjoint binary trees called the left subtree and the right subtree. A node contains information, the nature of which depends on the application.

The topmost node is usually called the root node, and the nodes at the bottom that have no other nodes beneath them are referred to as leaves. A subtree of the tree consists of a node in the tree along with all the nodes below it. Each node has no more than two nodes below it. That is why it is called a binary tree. The number at each node is greater (for strings in lexicographical order) than each of the numbers in the subtree that is below and to the left of it. On the other hand, each node is smaller than all the nodes in the subtree below and to the right of it. Thus we are able to search this tree very easily. Let say we have a tree of integer numbers. To find out whether a given number is in the tree we first compare with the value of the root node. If they are equal then the search is completed. If the given value is greater than the value of the root node we know that the given number cannot appear anywhere in the subtree to the left. Thus we search the subtree to the right. We repeat this search technique until we find the value or node is found. Instead of numbers we can also consider strings (names).

In our first program `tree1.cpp` we give a binary tree for `double`. In our second program `tree2.cpp` we give a template version.

8.6.2 A Simple Binary Tree

In our first program we consider a binary tree of `double`. The output of the member function `search` is 1 if the number is in the tree and 0 if it is not. The line

```
typedef double T;
```

can be changed to

```
typedef int T;
```

so that we have a tree of integers.

```
// tree1.cpp

#include <iostream.h>

typedef double T;

   // Declaration of class Tree
class Tree
{
   private:
      T root;
      int empty;
      Tree *left, *right;
   public:
      Tree();
      void insert(const T&);
      int search(const T&);
      void display();
};

 // Definition of class Tree
Tree::Tree() { empty = 1; }

void Tree::insert(const T& value)
{
   if(empty)            // short cut for (empty != 0)
   {
   root = value;
   left = new Tree;  right = new Tree;
   empty = 0;
   }
   else
   if(value <= root)
```

```
    left -> insert(value);  // insert at left
    else
    right -> insert(value); // insert at right
}

int Tree::search(const T& value)
{
    if(empty) return 0;
    else
    if(root == value) return 1;
    else
    if(root > value)  return left -> search(value);
    else
    return right -> search(value);
}

void Tree::display()
{
    if(empty) return;
    else
    {
    left -> display();
    cout << root << ' ';
    right -> display();
    }
}

void main()
{
    Tree B;
    B.insert(1.23);  B.insert(2.37);  B.insert(0.3);  B.insert(1.4);
    B.insert(7.1);   B.insert(0.1);   B.insert(-1.9); B.insert(8.3);
    B.display();
    cout << endl;

    int res1 = B.search(0.1);
    cout << "res1 = " << res1 << endl;  // 1 since 0.1 is in tree
    if(res1 == 1) cout << "number is in tree " << endl;
    else cout << "number is not in tree " << endl;

    int res2 = B.search(2.9);
    cout << "res2 = " << res2 << endl;  // 0 since 2.9 is not in tree
    if(res2 == 1) cout << "number is in tree " << endl;
    else cout << "number is not in tree " << endl;
}
```

8.6.3 Binary Trees and Templates

Here we give a template class for a binary tree.

```
// tree2.cpp

#include <iostream.h>
#include <assert.h>    // for assert

// Declaration of class Node
template <class T>
class Node
{
   private:
      T data;
      Node* left;
      Node* right;
   public:
      Node(const T&);   // constructor
      T getdata() const;
      friend class Tree<T>;
};

   // Definition of class Node
template <class T> Node<T>::Node(const T& d)
{
   data = d;
   left = NULL;  right = NULL;
}

template <class T> T Node<T>::getdata() const
{
   return data;
}

    // Declaration of class Tree
template <class T> class Tree
{
   private:
      Node<T>* root;
      void insertHelper(Node<T>**, const T&);
      void preOrderHelper(Node<T>*) const;
      void inOrderHelper(Node<T>*) const;
      void postOrderHelper(Node<T>*) const;
   public:
```

```
      Tree();
      void insert(const T&);
      void preOrderTraversal() const;
      void inOrderTraversal() const;
      void postOrderTraversal() const;
};

   // Definition of class Tree
template <class T> Tree<T>::Tree()
{ root = NULL; }

template <class T> void Tree<T>::insert(const T& value)
{
   insertHelper(&root, value);
}

template <class T>
void Tree<T>::insertHelper(Node<T>** p, const T& value)
{
   if(*p == NULL)
   {
   *p = new Node<T>(value);
   assert(*p != NULL);
   }
   else
   if(value < (*p)->data)
   insertHelper(&((*p) -> left), value);
   else
   if(value > (*p) -> data)
   insertHelper(&((*p) -> right),value);
   else
   cout << value << " dup" << endl;
}

template <class T> void Tree<T>::preOrderTraversal() const
{
   preOrderHelper(root);
}

template <class T>
void Tree<T>::preOrderHelper(Node<T>* p) const
{
   if(p != NULL)
   {
   cout << p -> data << ' ';
```

```
   preOrderHelper(p -> left);
   preOrderHelper(p -> right);
   }
}

template <class T>
void Tree<T>::inOrderTraversal() const
{
   inOrderHelper(root);
}

template <class T>
void Tree<T>::inOrderHelper(Node<T>* p) const
{
   if(p != NULL)
   {
      inOrderHelper(p -> left);
      cout << p -> data << ' ';
      inOrderHelper(p -> right);
   }
}

template <class T> void Tree<T>::postOrderTraversal() const
{
   postOrderHelper(root);
}

template <class T>
void Tree<T>::postOrderHelper(Node<T>* p) const
{
   if(p != NULL)
   {
   postOrderHelper(p -> left);
   postOrderHelper(p -> right);
   cout << p -> data << ' ';
   }
}

int main()
{
  Tree<char> charT;
  charT.insert('D');
  charT.insert('R');
  charT.insert('C');
  charT.insert('Z');
```

```
  charT.insert('X');
  charT.insert('U');

  cout << endl << "Preorder traversal" << endl;
  charT.preOrderTraversal();
  cout << endl << "Inorder traversal" << endl;
  charT.inOrderTraversal();
  cout << endl << "Postorder traversal" << endl;
  charT.postOrderTraversal();

  Tree<int> intT;
  intT.insert(2);
  intT.insert(7);
  intT.insert(10);
  intT.insert(-1);
  intT.insert(3);
  intT.insert(27);

  cout << endl << "Preorder traversal" << endl;
  intT.preOrderTraversal();
  cout << endl << "Inorder traversal" << endl;
  intT.inOrderTraversal();
  cout << endl << "Postorder traversal" << endl;
  intT.postOrderTraversal();

  Tree<char*> names;
  names.insert("willi");
  names.insert("otto");
  names.insert("carl");
  names.insert("franz");
  names.insert("ludwig");
  names.insert("jens");

  cout << endl << "Preorder traversal" << endl;
  names.preOrderTraversal();
  cout << endl << "Inorder traversal" << endl;
  names.inOrderTraversal();
  cout << endl << "Postorder traversal" << endl;
  names.postOrderTraversal();

  return 0;
}
```

8.7 A Set Class

The following program shows an implementation for a *finite set*. The implementation is based on a linked list. The operator + is overloaded for the union of two sets and the operator * for the intersection of two sets. The operator == is overloaded to compare two finite sets.

```
// set.cpp

#include <iostream.h>
#include <stdlib.h>

    // Declaration part class Set
template <class T> class Set
{
  public:
    Set();                      // default constructor
    Set(const T);               // constructor single element
    ~Set();                     // destructor
    Set(const Set <T>&);        // copy constructor
    Set<T>& operator = (const Set<T>&); // assignment operator
    int cardinality() const;    // number of elements
    friend Set<T> operator + (Set<T>&,Set<T>&); // union
    friend Set<T> operator * (Set<T>&,Set<T>&); // intersection
    friend int operator == (Set<T>&,Set<T>&);
    friend ostream& operator << (ostream&,Set<T>&);

     struct Cell {          // element of linked list
     T value;               // the value of the item in the set
     Cell* p_next;          // pointer to next Cell
  };

  struct Set_DESC {          // the Descriptor
     Cell  *p_root;          // root of list of values
     Cell  *p_last;          // last element
     int no_of_elements;     // number of elements in list
     int references;         // references to this item
 };

 protected:
   void first_element();          // iterate set first element
   void next_element();           // iterate move to next
   T current_element() const;     // iterate return current item
   int is_at_end() const;         // iterate at end
```

```
    void add_at_end(const T);   // add to list at end
    void release_storage();       // release sets storage
    void fail(const char[]) const; // give up

  private:
    Set_DESC* the_p_desc;    // description of set
    Cell *the_p_current;     // current element looked at
};

     // defintion part class Set
template <class T> Set<T>::Set()
{
   the_p_desc = new Set_DESC[1];
   if(the_p_desc == (Set_DESC*) 0)
   fail("Cannot create Set - descriptor");
   the_p_desc -> no_of_elements = 0;
   the_p_desc -> references = 1;
   the_p_desc -> p_root = (Cell*) 0;
   the_p_desc -> p_last = (Cell*) 0;
}

template <class T> Set<T>::Set(const T item)
{
   the_p_desc = new Set_DESC[1];
   if(the_p_desc == (Set_DESC*) 0)
   fail("Cannot create Set - descriptor");
   the_p_desc -> no_of_elements = 1;
   the_p_desc -> references = 1;
   the_p_desc -> p_root = new Cell[1];
   if(the_p_desc -> p_root == (Cell*) 0)
   fail("Cannot create Cell - descriptor");
   the_p_desc -> p_last = the_p_desc->p_root;
   the_p_desc -> p_root->value = item;
   the_p_desc -> p_root->p_next = (Cell*) 0;
}

template <class T> void Set<T>::release_storage()
{
   if(--(the_p_desc->references)==0)
   {
      Cell *p_active = the_p_desc->p_root;
      while(p_active != (Cell*) 0)
      {
      Cell *p_remove = p_active;
      p_active = p_active -> p_next;
```

```
      delete [] p_remove;
      }
      delete [] the_p_desc;
   }
}

template <class T> Set<T>::~Set() { release_storage(); }

template <class T> void Set<T>::fail(const char mes[]) const
{
   cout << "\n" << "Error in class : Set" << "\n";
   cout << mes << "\n";
   exit(-1);
}

template <class T> Set<T>::Set(const Set <T>& copy)
{
   copy.the_p_desc -> references++;
   the_p_desc = copy.the_p_desc;
   the_p_current = copy.the_p_desc -> p_root;
}

template <class T> int Set<T>::cardinality() const
{
   return the_p_desc -> no_of_elements;
}

template <class T> void Set <T>::first_element()
{
   the_p_current = the_p_desc -> p_root;
}

template <class T> void Set<T>::next_element()
{
   if(the_p_current != (Cell*) 0)
   the_p_current = the_p_current -> p_next;
}

template <class T> T Set <T>::current_element(void) const
{
   if(the_p_current == (Cell*) 0)
   {
   fail("Selecting item beyond end");
   }
   return the_p_current -> value;
```

```
}

template <class T> int Set<T>::is_at_end() const
{
   return the_p_current == (Cell*) 0;
}

template <class T> void Set <T>::add_at_end(const T item)
{
   Cell* p_add = new Cell[1];
   if(p_add == (Cell*) 0)
   fail("Cannot create new element");
   p_add -> value = item; p_add->p_next = (Cell*) 0;
   if(the_p_desc->p_root == (Cell*) 0)
   {
   the_p_desc->p_root = p_add;
   the_p_desc->p_last = p_add;
   } else {
   the_p_desc -> p_last -> p_next = p_add;
   the_p_desc -> p_last = p_add;
   }
   the_p_desc -> no_of_elements++;
}

template <class T> Set<T>& Set<T>::operator = (const Set<T>& s1)
{
   s1.the_p_desc->references++; // copied items ref count
   release_storage();           // storage for overwritten element
   the_p_desc = s1.the_p_desc;
   return *this;
}

template <class T> Set<T> operator + (Set<T>& s1,Set<T>& s2)
{
   Set<T> result;
   s1.first_element(); s2.first_element();
   while(!s1.is_at_end() || !s2.is_at_end())
   {
      if(s1.is_at_end())
      {
      result.add_at_end(s2.current_element());
      s2.next_element();
      } else
      if(s2.is_at_end())
      {
```

```
      result.add_at_end(s1.current_element());
      s1.next_element();
      } else
      if(s1.current_element() < s2.current_element())
      {
      result.add_at_end(s1.current_element());
      s1.next_element();
      } else
      if(s1.current_element() > s2.current_element())
      {
      result.add_at_end(s2.current_element());
      s2.next_element();
      } else
      if(s1.current_element() == s2.current_element())
      {
      result.add_at_end(s1.current_element());
      s1.next_element(); s2.next_element();
      }
   }
   return result;
}

template <class T> Set<T> operator * (Set<T>& s1,Set<T>& s2)
{
   int flag = 0;
   Set<T> intersection;
   s1.first_element();  s2.first_element();
   while(!s1.is_at_end())
   {
   while((!s2.is_at_end()) && (!flag))
   {
   if(s1.current_element() == s2.current_element())
   {
   intersection.add_at_end(s1.current_element());
   flag = 1;
   }
   s2.next_element();
   }
   flag = 0;
   s1.next_element();
   }
   return intersection;
}

template <class T> int operator == (Set<T>& s1, Set<T>& s2)
```

```
{
   s1.first_element(); s2.first_element();
   if(s1.cardinality() != s2.cardinality())
   return 0;
   if(s1.cardinality() == 0) return 1;
   while(!s1.is_at_end())
   {
   if(s1.current_element() != s2.current_element())
   return 0;
   s1.next_element(); s2.next_element();
   }
   return 1;
}

template <class T> ostream& operator << (ostream& s,Set<T>& s1)
{
   int i;
   s1.first_element(); s << "(";
   for(i=0; i < s1.cardinality(); i++ )
   {
   s << s1.current_element();
   s1.next_element();
   if(i != s1.cardinality()-1) s << ",";
   }
   s << ")";
   return s;
}

void main()
{
   Set<char> s1, s2, s3;
   Set<char> setA('X');  Set<char> setB('Y');
   Set<char> setC('Z');  Set<char> setD('Y');
   cout << setA << "\n";
   s1 = setA + setB;
   cout << "s1 = " << s1 << endl;
   s2 = setA + setB + setC + setD;
   cout << "s2 = " << s2 << endl;

   s3 = s1 * s2;
   cout << "s3 = " << s3 << endl;

   int size = s2.cardinality();
   cout << "size of s2 = " << size;
}
```

8.8 A Date and Month Class

8.8.1 Introduction

An abstract data type (which in C++ is called a `class`) defines not only the representation of the data (e.g. the day, month and year in the case of the `Date` example) but also the operations which may be performed on that data type. For example, incrementing the date is quite different from incrementing an integer. Incrementing 28th February 1994 should yield 1st March 1994 (1994 is not a leap year) and incrementing 31st December 1993 should yield 1st January 1994. Both the data representation of `Date` and the implementation details of the operations to be performed on `Date` should be hidden from programmers using this class (information hiding). Furthermore, in order to ensure integrity of the data fields, access to the data fields is restricted to the predefined methods and operators of the class. This is achieved in C++ by declaring the data fields private. Consider the following simplified definition of the interface for a `Date` class.

```
class Date
{
  private:
    int Day, Month, Year;
  public:
    void set(const int d,const int m,const int y);
    void get(int& d,int& m,int& y);
    void operator ++ ();
    int  operator < (const Date& d);
};
```

Following the philosophy of information hiding the data fields `Day`, `Month` and `Year` are declared private which makes them inaccessible from outside the class. The user has thus no direct access to the data fields. Instead these can be set with the function `set` which would check the validity of the date before setting the private data fields of the class equal to its input parameters. A further advantage of applying the concept of information hiding is that the designer of the date class could change the underlying data structure from the three integer fields `year`, `month` and `day` to a single long integer variable storing the number of days from a certain reference date. As we see later, such a change improves the efficiency of the class, both in terms of memory and in terms of computational speed (e.g. incrementing a date reduces to incrementing a single integer variable). Such a change would be completely transparent for a user — for all he knows the date could be stored in a string variable. The user interface for the class would remain unaltered.

The method `get` allows the user to retrieve the date in a format which is useful for him and the incrementation operator `++` allows the user to increment the date.

In the case where the underlying private data fields are `year`, `month` and `day`, the unary operator `++` would have to check whether it is an end of a month and if so reset `Day` to `1` and increment `Month` except if it is December in which case `Month` is set to `1` and `Year` is incremented. In all other cases it would simply increment `Day`. Furthermore, it has to make provision for leap years. All these implementation details are hidden from the user who simply increments `Date` like any other built-in variable. The binary operator $<$ compares two dates and returns `1` if the left hand side is an earlier date than the right hand side.

It is in general a good idea to strive for complete but minimal class interfaces. The class interface should be complete in order to allow the user of that class to do all the things one would like to do. In the case of the `Date` class it should allow the user to declare objects of the date class, to set and retrieve dates, to increment and decrement dates, to determine the time-span between two dates, to push the date onto an output stream and to read it from an input stream, to add two days, to specify the output format (i.e. whether the month should be written as a number or as a string) and to be able to make relational comparisons (i.e. `if (d1<d2) {...}`) between two dates. The class interface should also be minimal in order not to confuse the user and to decrease the amount of time required for the user to be able to use the class effectively. A minimal class interface usually has the additional advantage of reduced maintenance costs for the class.

A programmer could use the class `Date` in the following way

```
Date d1, d2;          // define 3 variables of type Date
d1.set(1,1,94);       // set d1 to 01-01-94
d2.set(7,8,5);        // set d2 to 07-08-95

for(Date d=d1; d<d2; ++d) { ... }
```

The `Date` class uses the `Month` class.

```
class Month
{
  private:
    int mnth;  // private integer holding the month
  public:
    Month();             // default constructor
    void thismonth(); // set the value of Month to
                      // month of current machine date
    int daysinmonth() const; // returns the no of days in that month
    Month& operator++ ();
};
```

The above class interface defines a constructor `Month()` which allows the user to create objects of type `Month` just as he would create variables of the built-in data types (e.g. `float`). The method `thismonth()` would set the object to the current month determined by the machine date and the method `daysinmonth` would return the number of days in that month. The interface also defines the incrementation operator for this class so that the user can increment a `Month` object. Note that the constructor `Month`, the method `thismonth` and the operator `++` alter the state of a `Month` object. The method `daysinmonth`, on the other hand, only returns the number of days in that month. It should leave the actual object unaltered and hence this method is declared constant.

Such a class could be used as follows:

```
void main()
{
  Month m1;
  m1.thismonth();
  cout << m1.daysinmonth();
  ++m1;
  cout << m1.daysinmonth();
}
```

Constructors enable the user to create objects of a certain class. They are methods which carry the name of the class and which have no return value. The simplified `Month` class above defined a constructor without arguments. Such a constructor is called the default constructor. Should we omit to define any constructor the helpful C++ compiler will write a default constructor with an empty body. Consider now the following extract from the full `Month` class listed at the end of this chapter

```
typedef enum outstyles { numeric, literal };
typedef enum months { unknown, January, February, March, April,
    May, June, July, August, September, October, November, December };

class Month
{
  private:
    int mnth;
    outstyles stl;
  public:
     // defining the various constructors
     Month() { mnth=unknown; stl=literal; };
     Month(const Month& m) { mnth=m.mnth; stl=m.stl; };
     Month(const months& m) { mnth=m; stl=literal; };
     Month(const int m);
}
```

The enumeration types `outstyles` and `months` are introduced to define a series of typed constants which will make the code more readable. The private data field `stl` holds the choice of output format — whether the `literal` (e.g. `21 September 1994`) or the `numeric` (e.g. `21/9/94`) representation of the month should be pushed onto output streams).

The class defines four different constructors — note that class member functions including constructors can be overloaded like any other function. The different constructors must be distinguishable by their parameter list. The first constructor has no arguments (it is called the default constructor). The body of this constructor is defined within the class definition block. Member functions whose body is defined within this block are by default defined as inline (as always with inline functions, this is a compiler request which the compiler might ignore if he has reason to do so – e.g. when the function body contains iterative statements). The default constructor initializes `mnth` to `unknown` and `stl` to `literal`. Should we omit to write any constructor for a class, the C++ compiler will write a default constructor. Default constructors are essential if we want to allow programmers to define arrays of objects of the class. The default constructor would be used if one declares

```
Month m;
```

The constructor `Month(Month& m)` is known as the copy constructor. It takes an object of the class `Month` as argument and makes a copy of that object. Hence

```
Month m2(m1);
```

would declare the `Month` object `m2` as a copy of `m1`. Should we omit to write a copy constructor, the C++ compiler will write one. Notice that the copy constructor written by the compiler will simply make a byte-for-byte copy of the data fields. Hence, if the class uses dynamic memory allocation, we will have to write the copy constructor. In the case of the `Month` class, the copy constructor written by the compiler would suffice.

The third constructor takes as its argument a variable of the enumeration type `months`. This argument is sent by reference and is declared constant since the constructor should not change this variable. This constructor sets the `Month` object to the month passed by the argument and initializes `stl` to `literal`

The fourth constructor receives as an argument an integer which the constructor treats as constant. Note that no body is defined in the class definition. The body could be defined as follows

```
Month::Month(const int m)
{
  stl = literal;
  if((m >=1) && (m <=12))
    mnth = (months)m;
  else
```

```
  {
    cerr << "*** ERROR *** : invalid month in Month(int)." << endl;
    mnth = unknown;
  }
};
```

Note that we have to use the scope resolution operator :: in order to specify to which class the method belongs. The stream `cerr` is the standard error stream which is usually the same as the standard output stream (i.e. the computer screen). A better way to do error handling is by throwing an exception. In the next chapter we introduce exception handling in C++.

Class methods give the user controlled access to the class members. In C++ one usually distinguishes between member functions and class operators. It is a good idea to distinguish from the start between member functions which only report about the state of an object and member functions which actually change the state of an object. For example the member function

```
iostyles Month::iostyle() const  { return stl; }
```

simply tells the user to which style the `stl` field is set (whether the month will be output in numeric or literal format). This member function does not change the state of the object and hence it is declared constant by appending `const` to the function header. On the other hand, the member function

```
void Month::setiostyle(const iostyle style)
{
stl = style;
}
```

sets the iostyle of a `Month` object. Hence it alters the state of that object — it cannot be declared constant. The above two methods could be used in the following form

```
Month m(January);
iostyles dstl = m.iostyle();
if(dstl != numeric)
d.setiostyle(numeric);
```

It is useful to define a method which returns the number of days in the current month. Since this method will not alter the state of the `Month` object it is declared constant. The new feature of the method `daysinmonth()` is that it declares a local static constant integer vector which holds the number of days in each month:

```
int Month::daysinmonth() const
{
static const int monthdays[] = {0,31,28,31,30,31,30,31,31,30,31,30,31};
return monthdays[mnth];
};
```

If the vector monthdays were not declared static, each object would have a copy of that vector. Since this vector should always be the same for all objects, there is no need to store it more than once. By declaring the vector static we make certain that there is only one copy of this vector in memory, irrespective of the number of Month objects declared. By declaring the vector constant we prevent them from being changed.

Now the above method returns the correct number of days for every month (it will return 0 if the field mnth is set to unknown, i.e. if the Month object was declared with the default constructor or declared with an invalid initialization parameter) except for leap-year Februaries. In order to return the correct number of days in this case we overload the daysinmonth method with a method of the same name, but receiving the year as an argument:

```
int Month::daysinmonth(const int year) const
{
  int nodays = daysinmonth();
  if((mnth == February) && (!(year%4)))
    ++nodays;
  return nodays;
};
```

To avoid code duplication (this is important also for maintenance considerations) we first call the default daysinmonth method and if the month is a leap-year February we add an additional day to the return value of the member function.

An abstract data type is not only defined by its data members, but also by the operations that can be performed on them. Incrementing a month is different from incrementing an integer (incrementing December should not yield 13, but January). When overloading operators one must bear in mind that in C++ one is restricted to using the predefined operators and also to the predefined precedence hierarchy. Hence, if we overload * and +, then * will still have higher precedence than + irrespective of the actual function they perform (which might have nothing to do with multiplication and addition). Furthermore, the syntax of an operator may not be changed in the sense that if a standard binary operator is overloaded, the new definition of the operator must still be binary (i.e. one cannot define a unary > or a binary ++).

To illustrate the formalism for operator overloading we choose one unary and one binary arithmetic operator (arbitrarily the incrementation operator ++ and the subtraction operator -). These are defined in the class interface as follows:

```
class Month
{
    ...
```

```
   Month& operator ++ ();
   Month  operator  - (const int nomonths) const;
   ...
}
```

Since the incrementation operator acts on the object itself, the operator method cannot be declared constant. The subtraction operator, on the other hand, leaves the actual object unchanged and returns the result of the operation in a new `Month` object. Hence this method is declared constant. The body of the incrementation operator is shown in the following listing

```
Month& Month::operator ++ ()
{
  if(mnth <= November)
    ++mnth;
  else
    mnth = January;
  return *this;
}
```

One would be tempted to define the incrementation operator such that it returns nothing (`void`). This would however disallow constructions like

```
  Month m1 = ++m2;
```

which is perfectly legal for the built-in types. In order to allow the programmer to use the new data type the same way he would be able to use the built in data types we let the incrementation operator return a reference to the object itself. For this purpose we use the pseudo-variable `this` which is a pointer to the current instance (to the object itself). By using the dereferencing operator `*` we are thus able to return the object itself. Recall that C++ defines both, prefix and postfix incrementation. In the case where we only define one of the two operators, it can be used in either prefix or postfix syntax. In the case where we want to distinguish between the two versions we can define them separately:

```
class Month
{
  ...
  operator++();        // prefix:  ++Month;
  operator++(int);     // postfix: Month++;
  ...
}
```

The only thing interesting about the body of the subtraction operator is the return statement:

```
Month Month::operator - (const int nomonths) const
{
  int m = ((mnth-nomonths)%12);
  if(m<=0) m+=12;
  return Month(m);
}
```

Note that Month(m) creates a new object by calling the constructor and that this new object is returned to the caller.

Should one omit to define an assignment operator, the C++ compiler defines one. In the case where static data members are used (as is the case for the Month class) the version supplied by the compiler will be accountable. Should the class, however, allocate memory dynamically, the assignment operator would simply copy the pointers, which would in most cases not be what is required. In such cases it is essential to define an own assignment operator. The assignment operator can be overloaded just like any other class method

```
class Month
{
  ...
  Month& operator = (const Month& m) { mnth = m.mnth; return *this; };
  Month& operator = (const int m);
  Month& operator = (const months& m) { mnth = m; return *this; };
  ...
}
```

Here we defined three forms for the assignment operator, allowing the user to assign one Month to another, assign a Month to an integer (a form of type conversion) and to assign a Month to a variable of the enumeration type months. The latter allows a programmer to formulate statements such as

```
Month m = January;
```

Note that the assignment operator again returns a reference to the object itself. This is essential to allow repetitative assignments:

```
Month m=m1=m2=m3=m4;
```

As for the operator ++, we force the assignment operator of our Month class to to behave in a similar way as it does for built-in data types for which such an assignment is legal.

In order to allow comparisons between Month objects we overloaded all the relational operators

```
==     !=     >     <     >=     <=+
```

They are straightforward binary operators and we just show one example here:

```
class Month
{
  ...
  int operator > (const Month& m) const
  { return (mnth >  m.mnth); };
}
```

To allow type conversions from `Month` to integer we define the type conversion operator

```
class Month
{
  ...
  operator int() { return mnth; };
}
```

Such type conversion operators are not restricted to the built-in data types. For example, if a class `linklist` is defined and a class `vector`, then the class `linklist` could define the type conversion operator `operator vector()` which would allow the user to do a type conversion from a linked list to a vector.

The one remaining problem is to allow the class to be pushed onto an output stream (e.g. written onto the screen) and to be able to extract the class from an input stream (e.g. initializing `Month` from the keyboard). To enable the C++ programmers to manipulate user-defined classes as easily as built-in data types, input and output streams are defined such that they can be easily modified to accommodate user-defined types. Access to input streams is obtained by overloading the stream operator `>>` (extract from)

```
istream& operator >> (istream& is,Month& m)
{
  int month;
  cin >> month;
  m = month;
  return is;
}
```

Here we extract an integer from the input stream and we use the assignment operator

```
Month& Month::operator = (const int) {...}
```

to convert the integer to the `Month` object `m`. We return a reference to the input stream object `is` to allow serial constructions of the form

```
cin >> month1 >> month2;
```

The output stream operator >> (push onto) is overloaded similarly:

```
ostream& operator << (ostream& os, Month& m)
{
  int month = m;
  switch(m.iostyle())
  {
  case numeric:
    { os << month; break; }
  case literal:
    static const char* const monthnames[] = {"unknown","January",
        "February","March","April","May","June","July","August",
        "September","October","November","December" };
    os << monthnames[month];
    break;
  }
  return os;
};
```

The first statement makes implicit use of the type conversion `operator int()` to convert the `Month` object to the integer `m`. In the case where the iostyle of the `Month` object is `numeric` it simply pushes the integer onto the output stream. In the case where the iostyle is `literal` we declare the static constant array of strings `monthnames` which carries the names of the months as its elements. Note again that declaring this array static assures that only one such array is stored, irrespective of the number of `Month` objects in use. Note that both the pointer to the array elements as well as the array elements themselves are declared constant. Finally we push the string with the relevant month name onto the output stream and as for the input stream we return a reference to the output stream object itself. This allows repetitive applications of the output stream operator <<:

```
Month m1 = January;
cout << "The month " << m1 << " has " << m1.daysinmonth()
     << " days." << endl;
```

A complete listing of the `Month` class is given at the end of the chapter.

The full `Date` introduces only a few new ideas (the main one being the concept of class constants and the way they are initialized). We look only at a few excerpts which illustrate these new ideas. The complete listing for the `Date` class is given again at the end of the chapter. The main interest for the reader should be the design ideas.

When prompted to define a date class one is tempted to define three data fields carrying the year, the month and the day in the month. In this example the date is stored in a single long integer variable which contains the number of days from a certain reference date. This idea is not original. Astronomers used the Julian date which counts the days sequentially from 1st January 4713BC with a correction of a few days in the middle ages. Here we have chosen the 1st January 1968 as reference date because it has the convenience that it is a leap year with the 1st of January falling on a Monday. Hence 1/1/68 is Day one and all other dates are simply stated as the number of days from that reference date (the number can be negative or positive).

Storing the date as a single long integer variable has several advantages over the intuitive approach of defining three private data members `year`, `month` and `day`. Firstly it requires less storage. More importantly, operations which would be used frequently (e.g. incrementing the date for compound interest calculations) are very easily and efficiently implemented (it simply involves incrementing an integer variable).

Since we abide to the practice of information hiding, the underlying data structure is unknown to the user who only interfaces with the class through the public member functions. For user input we define the function `set(day, month, year)` and `get(day, month, year)`.

The conversion to and from the underlying `long int` representation of `Date` is done behind the scenes by two privately declared member functions, `ToAbsolute(...)` and `ToConventional(...)`. Here is an excerpt of the `Date` class interface:

```
class Date
{
   private:
     const refyear;
     iostyles iostl;
     int absdate;
     int ToAbsolute(const int day, const Month& month,const int year);
     void ToConventional(int& day,Month& month,int& year) const;
  public:
      Date();
      Date(const int day,const Month& month,const int year);
      void today(); // setting Date to current date
      void SetDate(const int day,const Month& month,const int year);
      void GetDate(int& day,Month& month,int& year) const;
      iostyles iostyle() const { return iostl; };
      void setiostyle(const iostyles stl) {iostl = stl;}
      int DaysInYear() const;
      // incrementing and decrementing operators
      Date& operator ++();
      Date& operator --();
      // assignment operator
```

```
      Date& operator = (const Date& D);
      // adding(subtracting) nodays to(from) Date
      Date operator + (const long int nodays);
      Date operator - (const long int nodays);
      // defining the relational operators
      int operator < (const Date& D) const;
      ...
};

ostream& operator << (ostream& os,Date& D);
istream& operator >> (istream& is,Date& D);
```

The class constant `refyear` will be initialized by the class constructors to `1968`. The other private date members are `absdate` which holds the absolute date and `iostl` which holds the iostyle for `Date` (i.e. `numeric` for choosing output style `21/9/94` and `literal` for choosing output style 21 September 1994). Besides the two constructors there is a method `today()` which sets `Date` to the current machine date, methods to set and get the date in conventional format and a method which returns the number of days in that year (returning 365 or 366 depending on whether it is a leap year or not. This method is especially useful when calculating interest earned over a period). The increment and decrement operators are defined in such a way that the actual date is incremented or decremented and the addition and subtraction operators enable the user to add or subtract a number of days from a `Date`. Finally the relational operators are defined (comparing different dates) and stream access is obtained by overloading the `>>` and `<<` operators.

Class constants are initialized when the constructor is called. Hence different constructors can initialize a class constant differently. Consider the full definition of the constructor

```
Date(const int day,const Month& month,const int year)
     : refyear(1968)
     { absdate = ToAbsolute(day, month, year); iostl=literal; }
```

Note that the class constant `refyear` is not initialized in the constructor body, but in an initialization list. More than one data member can be initialized in that way, the different members being separated by commas. Note that any data members can be initialized in the initialization list. Constant data members and reference members, however, have to be initialized in the initialization list.

Some operations on the data field should not be visible to the user. In the case of the `Date` class this includes the conversion from the user-friendly date format to the format used internally. Since the user should not be exposed to these implementation details, the member functions performing the conversions are declared `private`:

```
class Date
```

```
{
  private:
   ...
  long int ToAbsolute(const int day,const Month& month,const int year);
  void ToConventional(int& day,Month& month,int& year) const;
   ...
}
```

Since `ToConventional` simply translates the current date into a user-friendly format without actually affecting the object, it is declared `const`. Both conversion functions make use of the method `Month::daysinmonth(const int year)`. The simplicity of the conversion routines is partially due to the choosing of the reference date such that day 1 of the absolute date falls on a Monday, January the first of a leap year:

```
int Date::ToAbsolute(const int day,const Month& month,const int year)
{
  int dyear = year-refyear;
  int days = dyear*365 + (dyear+3)/4;

  for(Month m=January; m<month; ++m)
    days += m.daysinmonth(year);

  days += day;
  return days;
}

void Date::ToConventional(int& day,Month& month,int& year) const
{
  day   = 1;
  month = January;
  year  = refyear;
  year += absdate/365.25;
  int absdays = absdate - (year-refyear)*365.25;

  while(absdays > month.daysinmonth(year))
  {
    absdays -= month.daysinmonth(year);
    ++month;
  }
  day = absdays;
}
```

Incrementing, decrementing, adding and subtracting are directly performed on `absdate`. For example

```
Date& operator ++ () { ++absdate; return *this; };
```

Note that the subtraction operator `-` is overloaded to either subtract a number of days from a date or to calculate the number of days between two dates:

```
Date Date::operator - (const long int nodays);
int Date::operator - (const Date D);
```

The function `today()` returns the current machine date. It uses the function `time()` and `localtime` which are part of the ANSI C standard. The function `time` gives the current time, in seconds, elapsed since `00:00:00` GMT, January 1, 1970, and stores that value in the location poined to by `timer`. The function `localtime` accepts the address of a value returned by `time` and returns a pointer to the structure of type `tm` containing the broken-down time. It corrects for the time zone and possible daylight saving time.

```
void Date::today()
{
  time_t long_time;
  time(&long_time);
  struct tm* tme;
  tme = localtime(&long_time);
  Month m(tme -> tm_mon);
  absdate = ToAbsolute(tme -> tm_mday,m+1,tme->tm_year);
};
```

8.8.2 Listing of Complete Month and Date Classes

```
// mydate.cpp

#include <iostream.h>
#include <iomanip.h>
#include <time.h>

typedef enum iostyles { numeric, literal };
typedef enum months { unknown, January, February, March, April, May,
       June, July, August, September, October, November,
       December };

class Month
{
  private:
    int mnth;
    iostyles stl;

  public:
    Month() {mnth=unknown; stl=literal;};
    Month(const months& m) { mnth=m; stl=literal; };
    Month(const int m);
    Month(const int m, iostyles style);

    int daysinmonth() const;
    int daysinmonth(const int year) const;
    int validmonth() const
    { return ((mnth>=January) && (mnth<=December)); };
    iostyles iostyle() const {return stl;};
    void setiostyle(const iostyles style) { stl = style; };

    // defining the assignment operators
    Month& operator = (const Month& m) {mnth = m.mnth; return *this;};
    Month& operator = (const int m);
    Month& operator = (const months& m) {mnth = m; return *this;};

    // defining the arithmetic operators
    Month& operator ++ ();
    Month& operator -- ();
    Month  operator  + (const int nomonths) const;
    Month  operator  - (const int nomonths) const;

    // defining the relational operators
```

```
    int operator >  (const Month& m)  const
    { return (mnth >  m.mnth); };
    int operator >= (const Month& m)  const
    { return (mnth >= m.mnth); };
    int operator <  (const Month& m)  const
    { return (mnth <  m.mnth); };
    int operator <= (const Month& m)  const
    { return (mnth <= m.mnth); };
    int operator == (const Month& m)  const
    { return (mnth == m.mnth); };
    int operator != (const Month& m)  const
    { return (mnth != m.mnth); };
    int operator >  (const months& m) const
    { return (mnth >  m); };
    int operator >= (const months& m) const
    { return (mnth >= m); };
    int operator <  (const months& m) const
    { return (mnth <  m); };
    int operator <= (const months& m) const
    { return (mnth <= m); };
    int operator == (const months& m) const
    { return (mnth == m); };
    int operator != (const months& m) const
    { return (mnth != m); };

    operator int() { return mnth; };

};  // end of class month

class Date
{
   private:
     const refyear;
     iostyles iostl;
     int absdate;

     int ToAbsolute(const int day, const Month& month,
                    const int year);
     void ToConventional(int& day, Month& month, int& year) const;

  public:
      // Default constructors
      Date() : refyear(1968)       // initializing class constant
      { today(); iostl=literal; };
```

```
// Constructor with initialization arguments
Date(const int day, const Month& month,const int year)
: refyear (1968)
{ absdate = ToAbsolute(day, month, year); iostl=literal; }

void today(); // setting Date to current date

void SetDate(const int day,const Month& month,
const int year);

void GetDate(int& day,Month& month,int& year) const
{ ToConventional(day, month, year); }

iostyles iostyle() const {return iostl;};
void setiostyle(const iostyles stl) {iostl = stl;}

int DaysInYear() const;

// incrementing and decrementing operators
Date& operator ++() { ++absdate; return *this; };
Date& operator --() { --absdate; return *this; };

// assignment operator
Date& operator = (const Date& D)
{ absdate = D.absdate; iostl = D.iostl; return *this; }

// adding(subtracting) nodays to(from) Date
Date operator+ (const long int nodays)
{ Date D(*this); D.absdate += nodays; return D; };
Date operator- (const long int nodays)
{ Date D(*this); D.absdate -= nodays; return D; };
long int operator- (const Date D)
{ return absdate-D.absdate; };

// defining the relational operators
int operator <  (const Date& D) const
{ return (absdate < D.absdate); }
int operator >  (const Date& D) const
{ return (absdate > D.absdate); }
int operator == (const Date& D) const
{ return (absdate == D.absdate); }
int operator != (const Date& D) const
{ return (absdate != D.absdate); }
int operator <= (const Date& D) const
{ return (absdate <= D.absdate); }
```

```
     int operator >= (const Date& D) const
     { return (absdate >= D.absdate); }
};

ostream& operator << (ostream& os, Date& D);
istream& operator >> (istream& is, Date& D);

//                    Implementation section
Month::Month(const int m)
{
  stl = literal;
  if((m >=1) && (m <=12))
  mnth = (months)m;
  else
  {
  cerr << "*** ERROR *** : invalid month in Month(int)." << endl;
  mnth = unknown;
  }
}

Month::Month(const int m,iostyles style)
{
  stl = style;
  if((m >=1) && (m <=12))
  mnth = (months)m;
  else
  {
  cerr << "*** ERROR *** : invalid month in Month(int)." << endl;
  mnth = unknown;
  }
}

Month& Month::operator = (const int m)
{
  if((m >=1) && (m <=12))
  mnth = (months)m;
  else
  {
  mnth = unknown;
  cerr << "*** ERROR *** : invalid month in Month(int)." << endl;
  }
  return *this;
}

int Month::daysinmonth() const
```

```
{
  static const int monthdays[] =
  { 0,31,28,31,30,31,30,31,31,30,31,30,31 };
  return monthdays[mnth];
}

int Month::daysinmonth(const int year) const
{
  int nodays = daysinmonth();
  if ((mnth == February) && (!(year%4)))
  ++nodays;
  return nodays;
}

Month& Month::operator ++ ()
{
  if(mnth <= November)
    ++mnth;
  else
    mnth = January;
  return *this;
}

Month& Month::operator -- ()
{
  if (mnth >= February)
  --mnth;
  else
  mnth = December;
  return *this;
}

Month Month::operator + (const int nomonths) const
{
  int m = ((mnth+nomonths)%12);
  if (m<=0) m+=12;
  return Month(m);
}

Month Month::operator - (const int nomonths) const
{
  int m = ((mnth-nomonths)%12);
  if(m<=0) m+=12;
  return Month(m);
}
```

```
ostream& operator << (ostream& os,Month& m)
{
  int month = m;
  switch(m.iostyle())
  {
    case numeric:
    { os << month; break; }
    case literal:
      static const char* const monthnames[] = {"unknown","January",
      "February","March","April","May","June",
      "July","August","September","October",
      "November","December"};
       os << monthnames[month];
       break;
  }
  return os;
};

istream& operator >> (istream& is,Month& m)
{
  int month;
  cin >> month;
  m = month;
  return is;
}

void Date::SetDate(const int day,const Month& month,const int year)
{ if((day<1) || (day>month.daysinmonth(year)))
  cerr << "*** ERROR *** invalid date" << endl;
absdate = ToAbsolute(day,month,year); iostl=literal; }

void Date::today()
{
  time_t long_time;
  time(&long_time);
  struct tm *tme;
  tme = localtime(&long_time);
  Month m(tme -> tm_mon);
  absdate = ToAbsolute(tme -> tm_mday, m+1, tme -> tm_year);
}

int Date::ToAbsolute(const int day,const Month& month,
                     const int year)
{
```

```
  int dyear = year-refyear;
  long int days = dyear*365 + (dyear+3)/4;

  for(Month m=January; m<month; ++m)
    days += m.daysinmonth(year);

  days+=day;

  return days;
}

void Date::ToConventional(int& day,Month& month,int& year) const
{
  day   = 1;
  month = January;
  year  = refyear;
  year += absdate/365.25;

  int absdays = absdate - (year-refyear)*365.25;

  while(absdays > month.daysinmonth(year))
  {
    absdays -= month.daysinmonth(year);
    ++month;
  }
  day = absdays;
}

int Date::DaysInYear() const
{
  int day, year;
  Month month;

  GetDate(day,month,year);

  if(year%4)
  return 365;
  else
  return 366;
}

istream& operator >> (istream& is,Date& D)
{
  int day, year;
  Month month;
```

```
  is >> day >> month >> year;

  D.SetDate(day, month, year);

  return is;
}

ostream& operator << (ostream& os,Date& D)
{
  int   year;
  Month month;
  int   day;

  D.GetDate(day,month,year);

  month.setiostyle(D.iostyle());

  if(D.iostyle() == literal)
    os << day << ' ' << month << ' ' << year;
  else
    os << day << '/' << month << '/' << year%100;

  return os;
}

// A small program to test the class Month and class Date
int main()
{
  Month m1(March), m2(September);
  cout << "(" << m1 << " > " << m2 << ") =  " << (m1>m2) << endl;

  cout << endl << "Testing for loop" << endl
       << "----------------" << endl ;

  for(Month m3 = m1; m3<=m2; ++m3)
    cout << m3 << endl;

  Month m4 = ++m3;
  cout  << endl << "m4 = " << m4 << endl << endl;

  cout << "m3 = " << m3 << endl;
  cout << "m3+11 = " << m3+11 << endl;
  cout << "m3-11 = " << m3-11 << endl;
```

```
  int k = m3;
  cout << "type conversion to integer: k = m3 = " << k << endl;

  Date d1;

  cout << "d1 = " << d1 << endl;

  cout << "d1 + 62 days = " << d1+62 << endl << endl;

  Date d2(25,February,1892);
  Date d3(3,March,1892);

  for(Date d4=d2; d4<=d3; ++d4)
  cout << d4 << endl;
  cout << endl;

  d2.SetDate(25,February,1994);
  cout << d2 << " - " << d3 << " = " << d2-d3 << " days."
       << endl << endl;
  d3.SetDate(3,March,1994);
  d2.setiostyle(numeric);

  for(d4=d2; d4<=d3; ++d4)
  cout << d4 << endl;

  cout << "Enter Date: ";
  // for example: 25 2 2000
  cin  >> d4;
  cout << "You entered the date " << d4 << endl;

  return 0;
}
```

Chapter 9

File Manipulations

9.1 Introduction

In Chapter 2 we considered simple C++ I/O streams. The compiler uses a header for the definitions of I/O stream classes, the class methods and variables. Using I/O stream manipulators and member functions, the programs performed a variety of I/O operations. This chapter examines C and C++ file I/O operations in detail. Using file streams the program can perform I/O with insertion and extraction operators and member functions. The following topics are discussed for both C and C++:

- How to open and close a file
- Which class types the header file `fstream.h` defines
- After a file is opened, how one can perform output and input operations
- How to perform a formatted file output
- How to test each operation's success, so that, should an operation fail, the program can then detect and respond to the error
- What binary file operations are and what the programs need to read or write structures, arrays, or even floating-point values to files
- Sequential file operations and random file operations
- How to position file pointers for random input and output operations

In chapters 2 to 8 our programs performed extensive I/O operations to the I/O streams

```
cin    cout
```

To use these I/O streams, the programs simply referenced the stream names. When one uses these streams, C++ associates the streams to the screen and keyboard automatically, behind the scenes. When the programs read information from or write information to a file, however, they must specifically open the desired file, and one must specify whether one want to read from an input file or write to an output file. To understand file streams better for C++, we have to examine the header file `fstream.h`. Within the file one finds class definitions for

```
ifstream   ofstream   fstream
```

When the programs open files, one creates objects using these class types, based on the type of I/O operation one needs to perform. For example, to perform file input operations, one uses the `ifstream` class. Likewise, for file output operations one uses the `ofstream` class. Finally, in some cases the programs have to read or write the same file. In such cases, one uses the `fstream` class. The `ifstream` class is derived from the `istream` and `fstreambase` classes, and `istrstream` is derived from `istream` and `strstreambase`. The C++ language provides for virtual inheritence to avoid multiple declarations. This means that all the members (data and functions) of

```
iostream  istream  ostream  fstreambase  ios
```

are part of objects of the `fstream` class.

The input and output (I/O) routines of the standard C library allow us to read and write data to and from files an devices. The function `fopen` opens a specified file for stream I/O. It returns a pointer to the open file if successful; `NULL` if not. The function `fclose` closes an open file. It returns 0 if the stream is successfully closed; `EOF` if not.

In C	**In C++**	**Explanation**
`char c = getchar();`	`char c; cin >> c;`	Read the next character from the standard input stream into the character variable `c`.
`FILE *in;`	`ifstream in;`	Declare a stream (input)
`in=fopen("i.txt","r");`	`in.open("i.txt");`	Associate the file `i.txt` with the stream `in`.
`c = getc(in);`	`in >> c;`	Read the next character from the stream `in` into `c`.
`putchar(c);`	`cout << c;`	Write the character contained in `c` to the standard output stream.
`FILE *out;`	`ofstream out;`	Declare a stream `out` (output)
`out = fopen("o.txt","w");`	`out.open("o.txt");`	Associate the file `o.txt` with the stream `out`.
`putc(c,out);`	`out << c;`	Write the character contained in `c` to the stream `out`.
`fprint(out,"Message");`	`out << "Message";`	Write the `"Message"` to the stream `out`.

9.2 C Style File Manipulations

9.2.1 Output to File

In the following program we write the output to a file. The function `fopen` opens a file and associates a stream with it. `fopen` returns a pointer to be used to identify the stream in subsequent operations. We consider the case that the output is a string, an integer and an array. The function `fclose` closes the file.

```
// output.cpp

#include <iostream.h>
#include <stdio.h>
#include <string.h>

void main()
{
   char *str1 = NULL;
   str1 = new char[6];
   strcpy(str1,"Willi");

   FILE *stream;
   stream = fopen("result.str","w+");
   fprintf(stream,"%s",str1);
   fclose(stream);

   int a = 4; int b = 3; int c = a + b;
   stream = fopen("result.num","w+");
   fprintf(stream,"%d",c);
   fclose(stream);

   float x[10];
   x[0] = 0.618;
   for(int j = 0; j<9; j++)
   {
   x[j+1] = 4.0*x[j]*(1.0 - x[j]);
   }
   stream = fopen("array.dat","w+");
   for (j = 0; j<10; j++)
   {
   fprintf(stream,"%f\n",x[j]);
   }
   fclose(stream);
}
```

9.2.2 Reading String from File

The following program reads a string from a file and displays them on the screen. The file name is `solution.html`.

```
// readfile.cpp
#include <iostream.h>
#include <stdio.h>
#include <stdlib.h>    // for exit

int check(unsigned long k,unsigned long* keys,int length)
{
   int count = 0;
   while(count < length)
   {
   if(k == keys[count])  return 1;
   count++;
   }
   return 0;
}

void main()
{
   int length = 4;
   unsigned long* keys = NULL;
   keys = new unsigned long [length];
   keys[0] = 1233; keys[1] = 3456; keys[2] = 4198; keys[3] = 5678;
   unsigned long k;
   cout << "enter your key: ";   cin >> k;
   int result = check(k,keys,length);

   if(result == 1)
   {
   FILE* fptr;
   char str[81];
   fptr = fopen("c:/cppcourse/solution.html", "r");
   if(fptr == NULL) { cout << "Error: cannot open file";  exit(0); }
   while(fgets(str,80,fptr) != NULL)
   cout << str;
   fclose(fptr);
   }   // end if
   if(result == 0)
   cout << "wrong key";
}
```

9.2.3 Word Count in File

In the following program we count the words in a file. The program is a C version. Let say the file is named `myfile.txt`. First we generate the file `wordcount.exe`. Then at the command prompt we type:

```
wordcount myfile.txt
```

```
// wordcount.cpp

#include <stdio.h>      // for getc(),
#include <stdlib.h>     // for exit()

void main(int argc,char *argv[])
{
   FILE *fptr;
   int ch, string[81];
   int white = 1;           // white space flag
   int count = 0;           // word count
   if(argc !=2)             // check number of arguments
      {
      printf("Format: C>type2 filename"); exit(0);
      }
      if((fptr=fopen(argv[1],"r")) == NULL)  // open file
      {
      printf("Can't open file %s.", argv[1]); exit(0);
      }
      while((ch = getc(fptr)) != EOF)  // get char from file
      switch(ch)
      {                              // if space, tab or
      case ' ':                      // newline, set flag
      case '\t':
      case '\n': white++; break;
      default: if(white)
      { white = 0; count++; }
      }
      fclose(fptr);                           // close file
      printf("File %s contains %d words.", argv[1],count);
 }
```

9.2.4 File Searching Program

The following program `fsearch.cpp` searches a file for a specified string. First we generate the file `fsearch.exe`. If for instance the name of the file we want to search is `data.dat` and the string to search for is `union`. Then at command prompt we type:

```
fsearch data.dat union
```

The program uses the functions `fgets` and `fputs`. The command

```
char *fgets(char *s,int n,FILE *stream);
```

reads characters from `stream` into the string `s`. The function stops reading when it reads either $n-1$ characters or a newline character, whichever comes first. The function

```
char *fputs(const char *s,FILE *stream);
```

copies the null-terminated string s to the given output stream. It does not append a newline character, and the terminating null character is not copied. The program is a C version, except that we use `cout`.

```
// fsearch.cpp
// searches a file for a specified string

#include <stdio.h>
#include <iostream.h>
#include <string.h>    // for strstr()
#include <stdlib.h>    // for exit()

const int LINELEN = 80;

void checkline(char *line,char *str,int num)
{
   if((strstr(line, str)) != NULL)
   {
   printf("line: %s pattern: %s\n line number: %d\n",line,str,num);
   cout << "\n";
   }
}
```

```
int main(int argc,char *argv[])
{
   FILE *fp;                    // file pointer
   int count;                   // the line counter
   char line[LINELEN + 1];      // read input strings

   if((argc < 2) || (argc > 3))
   {     // check arguments
   cout << "\n" << "Invalid number of arguments";
   cout << "\n" <<  "Try: fsearch filename string";
   exit(1);
   }

   if(!(fp = fopen(argv[1], "r")))
   {  // open input file
   printf("\nFile cannot be opened: %s", argv[1]);
   exit(1);
   }

   count = 1;
   while(!feof(fp))
   {      // read input file
   fgets(line,80,fp);        // read a string
   checkline(line,argv[2],count);
   count++;
   }
   cout << "\n";
   printf("%d line(s) read in file %s\n",count, argv[1]);

   fclose(fp);

   return 0;
}
```

9.2.5 Counting Characters in a File

The program charcount.cpp counts the characters in a file. Let data.dat be the file name, where we want to count the characters. We use a command-line argument to obtain the name of the file to be examined. We check the number of arguments of the main function. If there are two, then the second one, arg[1], is the name of the file. After generating the execute file charcount.exe of the program, we type at the command prompt:

```
charcount data.dat
```

The second argument of the function main is the file name data.dat.

```
// charcount.cpp

#include <stdio.h>   // for printf(), fopen(), fclose(), getc()
#include <stdlib.h>  // for exit()

int main(int argc, char *argv[])
{
   FILE *fptr;
   char string[81];
   int count = 0;
   if(argc != 2)                     // check number of arguments
      {
      printf("Format: C>charcount filename");
      exit(0);
      }
   if((fptr=fopen(argv[1], "r")) == NULL)  // open file
      {
      printf("Can't open file %s.", argv[1]);
      exit(0);
      }
   while(getc(fptr) != EOF)          // get char from file
   count++;                          // count it

   fclose(fptr);                     // close file
   printf("File %s contains %d characters.",argv[1],count);

   return 0;
}
```

The second program counts the characters in a phrase typed in. The line

```
(getche() != 'r')
```

checks whether [RETURN] is pressed.

```
// charcnt.cpp

#include <iostream.h>
#include <conio.h>

int main()
{
   int n = 0;

   cout << " enter a phrase: ";

   while(getche() != '\r')
   n++;

   cout << "\n" << " number of characters = " << n;

   return 0;
}
```

The following program counts the number of characters in a file. The file name in the program is `article.tex`. The function `getc` is a macro that returns the next character on the given input stream and increments the stream's file pointer to point to the next character.

```
// charfile.cpp

#include <iostream.h>
#include <stdlib.h>
#include <stdio.h>

int main()
{
   FILE *in;

   char string[81];
   int n = 0;

   if((in=fopen("article.tex", "r"))==NULL)
   {
   cout << " cannot open intput file " << "\n";
   exit(0);
   }
   while(getc(in) !=EOF)
   n++;

   fclose(in);

   cout << "File contains " << n << " characters";

   return 0;
}
```

9.2.6 Class and Standard Output

The member function `show()` in the `class Point` prints out the coordinates of a point. The default output file is declared as `stdout`. If no argument is given when the function `show()` is invoked, the value of `f` will default to the standard output device, `stdout`. Thus we can decide whether we want to print to the screen or to a file.

```
// standard.cpp

#include <stdio.h>       // for file manipulation
#include <iostream.h>    // for cout
#include <math.h>        // for sqrt


  // Declaration of class Point
class Point
{
   private:
      double x, y;
   public:
      Point();                    // default constructor
      Point(double,double);       // constructor two arguments
      Point(const Point&);        // copy constructor
      void show(FILE *f=stdout); // display point C style
      Point& operator = (const Point&); // overloading =
      Point operator + (const Point&);  // overloading +
      friend double length(const Point*,const Point*); // length
};

  // Definition of class Point
Point::Point() { x = 0.0; y = 0.0; }

Point::Point(double a, double b) { x = a;  y = b; }

Point::Point(const Point& arg) { x = arg.x; y = arg.y; }

void Point::show(FILE *f)
{
   fprintf(f, "Point is (%lf,%lf)\n", x, y);
}

Point Point::operator + (const Point& u)   // overloading +
{
   Point t;
```

```
   t.x = x + u.x;   t.y = y + u.y;
   return t;
}

Point& Point::operator = (const Point& v)   // overloading =
{
   if(this != &v)  {  // check if left side is same as right
     x = v.x;  y = v.y;
   }
   return *this;
}

double length(const Point *p0, const Point *p1)
{
   double r = (p0->x-p1->x)*(p0->x-p1->x)+(p0->y-p1->y)*(p0->y-p1->y);
   return sqrt(r);
}

int main()
{
   Point q1, q2(2.5,0.7);
   Point q3 = q2;
   FILE* fout;
   fout = fopen("result.dat","w");    // C style "w" write only

   q2.show();        // on screen: Point is (2.500000,0.700000)
   q3.show(fout);  // in file result.dat: Point is (2.500000,0.700000)

   Point q4(3.7,-0.9);
   Point q5 = q4 + q2;
   q5.show();        // on screen: Point is (6.200000,-0.200000)

   double dist = length(&q2,&q4);
   cout << "distance = " << dist << endl;   // 2

   Point q6(2.3,7.9), q7(3.4,8.9), q8(10.1,-9.3);
   Point q9 = q6 + q7 + q8;
   q9.show();        // on screen: Point is (15.800000,7.500000)
   cout << endl << endl;

   fclose(fout);

   return 0;
}
```

9.2.7 The Functions fseek, fread, fwrite

The function `fseek` repositions a file pointer on a stream. The function `fread` reads data from a stream. The function `fwrite` writes to a stream. The function `fopen` opens a stream and the function `fclose` closes a stream. These function are included in the header file `stdio.h`. The function `int remove(const char* filename)` deletes the file specified by filename. If the file is open, be sure to close it before removing it. The class `fVektor` defines a vector, whose contents is in a file The class `FilePos` serves to find a specific position. The program deals with huge vectors transfered to file.

```
// FVEKTOR.CPP

#include <stdio.h>    // for fseek, fread, fwrite, fopen, fclose
#include <string.h>   // for strcpy
#include <iostream.h> // for cout

// forward declaration
template <class T> class fVektor;

 // Declaration of class FilePos
template <class T> class FilePos
{
   protected:
     friend class fVektor<T>;
     FILE  *Filepointer;
     long Position;
   public:
     operator T ();           // type conversion
    // FilePos to type T
     void operator = (T i); // overload assignment
     friend ostream& operator << (ostream&, FilePos<T>);
};

// Implementation of class FilePos
// type conversion abstract data type FilePos to type T
template <class T> FilePos<T>::operator T ()
{
  T n;
  fseek(Filepointer, Position, SEEK_SET);
  fread(&n, sizeof(T), 1, Filepointer);
  return n;
}
// overloading = write filepointer into file
template <class T> void FilePos<T>::operator = (T i)
```

```
{
  fseek(Filepointer, Position, SEEK_SET);
  fwrite(&i, sizeof(T), 1, Filepointer);
}

// overloading the operator <<
template <class T> ostream& operator << (ostream& s, FilePos<T> p)
{
  return s << T (p);  // here the type conversion is called
}

      // Declaration of class fVektor
template <class T> class fVektor
{
    public:
      fVektor(long i0,long i1,char *Datei);
      ~fVektor();
      FilePos<T> operator [] (long i); // overload []
    protected:
      FILE *Filepointer;
      char Filename[10];
      long Size, Basis;
};

// Implementation of class fVektor
template <class T> fVektor<T>::fVektor(long i0,long i1,char *Datei)
{
   Size = i1 - i0 + 1;
   Basis = i0;
   strcpy(Filename,Datei);
   Filepointer = fopen(Datei,"w+"); // create a new file for update
}

// close and delete file
template <class T> fVektor<T>::~fVektor()
{
  fclose(Filepointer);
  remove(Filename);
  cout << "destructor invoked: " << endl;
}

template <class T> FilePos<T> fVektor<T>::operator[] (long i)
{
   FilePos<T> p;
   p.Filepointer = Filepointer;
```

```
   p.Position = (i - Basis)*sizeof(T);
   return p;
}

void main ()
{
   long i;
   fVektor<int> A(5000,10000,"temp1");
   for(i=5000; i < 10000; i++) A[i] = 2*i;
   for(i=5000; i < 10000; i+=1000) cout << A[i] << " ";
   cout << endl;

   fVektor<char> B(10,30,"temp2");
   for(i=10; i < 30; i++) B[i] = 'A' + i;
   for(i=10; i < 30; i++) cout << B[i] << " ";
   cout << endl;

   fVektor<char*> C(3,6,"temp3");
   C[3] = "willi"; C[4] = "sue"; C[5] = "helen";
   for(i=3;i < 6; i++)
   cout << C[i] << " ";
   cout << endl << endl;
}
```

The output is:

```
10000 12000 14000 16000 18000
K L M N O P Q R S T U V W X Y Z [ \ ] ^
willi sue helen
destructor invoked
destructor invoked
destructor invoked
```

9.2.8 The Function fscanf

The function `fscanf`

```
int fscanf(FILE* stream,const char* format[,address,...]);
```

scans and formats input from a stream. It scans a series of input fields, one character at the time, reading from a stream. Then each field is formatted according to a format specifier passed to `fscanf` in the fromat string pointed to by format. If `fscanf` attempts to read at end-of-file, the return value is `EOF`. If no fields were stored, the return value is 0.

The following program shows an application of `fscanf`. The data file `data1.dat` is given by

```
2345 45.3 ute
```

The data file `data2.dat` is given by

```
2345 45.3 ute
34123 -7.12 willi
```

```
// readin.cpp

#include <stdio.h>
#include <string.h>
#include <iostream.h>

int main()
{

  FILE* fptr;
  fptr = fopen("data1.dat","r");

  int i;
  float x;
  char* str = NULL;
  str = new char [4];

  fscanf(fptr,"%d%f%s",&i,&x,str);
  cout << "i = " << i << endl;
  cout << "x = " << x << endl;
  cout << "str = " << str << endl;

  fclose(fptr);
```

```
  delete str;

  fptr = fopen("data2.dat","r");
  int a[2];
  float y[2];
  char astr[2][10];

  int j;
  for(j=0; j<2; j++)
  {
  fscanf(fptr,"%d%f%s",&a[j],&y[j],astr[j]);
  }
  cout << "a[0] = " << a[0] << endl;
  cout << "a[1] = " << a[1] << endl;
  cout << "y[0] = " << y[0] << endl;
  cout << "y[1] = " << y[1] << endl;
  cout << "astr[0] = " << astr[0] << endl;
  cout << "astr[1] = " << astr[1] << endl;

  fclose(fptr);

  return 0;
}
```

9.3 C++ Style File Manipulation

9.3.1 Output to File

In the following program we write the output to a file using C++ style. The program creates an object of type `ofstream`, which is defined within the header file `fstream.h`.

```
// fileop.cpp

#include <fstream.h>

int main()
{
   ofstream work1;        // work1 is object of type ostream

   work1.open("sum.dat"); // open the desired file named sum.dat
   int a = 2;
   int b = 3;
   int c = a + b;
   work1 << c;            // perform the output operation
   work1.close();

   ofstream work2;        // close the file sum.dat

   work2.open("values.dat");

   int array[4] = { 2, 3, 5, 1 };

   work2 << "working hours of Joe" << endl;
   for(int i=0;i<4; i++)
   work2 << "array[" << i << "] = " << array[i] << endl;

   work2.close();

   return 0;
}
```

9.3.2 Reading String from File

We reading a string from a file. The function `getline()` is a member of each input stream class. Its prototypes are

```
istream& getline(char* buf,streamsize num);
istream& getline(char* buf,streamsize num,char delim);
```

The first form reads characters into the array pointed to by `buf` until either `num-1` characters have been read, a newline character has been found, or the end of the file has been encountered. The array pointed to by `buf` will be null-terminated by `getline()`. If the newline character is encountered in the input stream, it is extracted, but is not put into `buf`. The second form reads characters into the array pointed to by `buf` until either `num-1` characters have been read, the character specified by `delim` has been found, or the end of file has been encountered. The array pointed to by `buf` will be null-terminated by `getline()`. If the delimiter character is encountered in the input stream, it is extracted, but is not put into `buf`.

```
// readf2.cpp

#include <fstream.h>

int main()
{
   char string[81];

   ifstream data("textfile.txt");

   data.getline(string, sizeof(string));

   cout << "string = " << string << endl;

   data.close();

   return 0;
}
```

9.3.3 String Replacement in File

The following program illustrates simple file input and output as well as simple strings. We use the standard C++ strings which are defined as arrays of characters with a terminating NULL character. Hence the string variable inputfilename is defined as an array of 60 characters. The string can be no longer than that length with the actual length determined by the position of the NULL character in that array of characters.

In the following program the contents of a file are searched and all occurrences of oldstr are replaced with the string newstr with the result written to a new file.

```
// REPLACE.CPP

#include <stdlib.h>                 // for exit()
#include <string.h>                 // for strlen()
#include <fstream.h>                // for fileIO
#include <iostream.h>               // for console IO

void main()
{
  char inputfilename[60];
  cout << "Enter name of input file: ";
  cin  >> inputfilename;

  ifstream infile (inputfilename);  // opening input stream
  if (!infile)                      // fromfile inputfilename
  {
     cout << "*** ERROR *** :  cannnot open " << inputfilename
  << endl;
     exit(1);  // Flushes output buffers, closes files,
  };           // aborts program

  char outputfilename[60];
  cout << "Enter name of output file: ";
  cin  >> outputfilename;

  ofstream outfile (outputfilename); // open output file stream

  char oldstr[60], newstr[60];

  cout << "Enter string to be replaced throughout file: ";
  cin  >> oldstr;
  cout << "Enter replacement string:                    ";
```

```
 cin  >> newstr;

 char c;  int k=0;  long int count=0;

 while(!infile.eof())   // while not end of input file
 {
   infile.get(c);  // read next character from input file

   if(c==oldstr[k])      // if character conforms to next
   {                     // character of string to be replaced
   ++k;                  // increment pointer which points to
 // number of characters which agree.
   if(k==strlen(oldstr)) // if entire string agrees with
   {                       // string to be replaced, write
   outfile << newstr;      // replacement string to output
   k = 0;                  // file & reset substring pointer.
   ++count;                // counting no of replacements.
   }
   }
   else                    // if next character does not agree:
   {
   for(int kk=0; kk<k; ++kk) // first write substring which
   outfile << oldstr[kk];    // did agree,
   outfile << c;             // then the char which did not agree
   k=0;                      // and reset substring pointer
   }
  };
  cout << "The string <" << oldstr << "> has been replaced by <"
       << newstr << "> " << count << " times." << endl;

  infile.close();
  outfile.close();
};
```

An input file is opened by creating an object of class `ifstream` with the file name as parameter. The statement

```
if(!infile) ...
```

checks whether the file `infile` was opened successfully. We use the stream method `get()` instead of the `>>` operator, since the latter filters out spaces, carriage return and line feed characters. When the files are no longer required they are closed via the stream method `close()`.

9.3.4 Reading and Writing of File

The following code will write to the file `mas.txt` and then read back the data displaying it on the user's terminal. *Whitespace* is the collective name given to: spaces (blanks), horizontal and vertical tabs, newline characters and comments. The manipulator

```
resetiosflags(long f)
```

clears the format bits specified by `f`. Whitespace skipping is controlled by the `ios::skipws` flag in the format state's enumeration.

```
// whitespace.cpp

#include <fstream.h>    // for ifstream, ofstream
#include <iomanip.h>
#include <stdlib.h>    // for exit()

void main()
{
   ofstream sink;    // output stream
   ifstream source;  // input stream
   char c;

   sink.open("mas.txt");  // connect to mas.txt
   if(!sink)              // failed to create
   {
   cerr << "Failed to create mas.txt" << endl;
   exit(-1);
   }
   sink << "Hello world" << endl;
   sink.close();           // close stream

   source.open("mas.txt"); // connect to mas.txt
   if(!source)             // failed to open
   {
   cerr << "Failed to open mas.txt" << endl;
   exit(-2);
   }
   source >> resetiosflags(ios::skipws); // read white space
   while(source >> c, !source.eof())    // copy mas.txt to cout
   {
   cout << c;
   }
   source.close();   // close stream
}
```

The following file manipulation program does not run under Windows NT and Visual C++ since the compiler does recognize the `m` in `strstream`. The command `ios::in` opens for input and `ios::nocreate` tests whether the file exists. If the file does not exist open fails.

```
// iofile.cpp

#include <iostream.h>
#include <fstream.h>
#include <strstream.h>
#include <stdlib.h>
#include <string.h>

int main()
{
   ifstream fin;

   char name[30], outputstr[256];

   ostrstream outs(outputstr, sizeof(outputstr));

   double value;

   fin.open("names.dat", ios::in | ios::nocreate);
   if(!fin)
   {
   cerr << "could not open file names.dat " << endl;
   exit(1);
   }

   while(fin >> name)
   {
   fin >> value;
   outs << name << " = " << value << " ";
   }

   outs << ends;
   cout << outputstr << endl;

   fin.close();

   return 0;
 }
```

The following program counts the number of { and }, the number of (and) and the number of [and] in a C++ program. If the number does not compensate the output of the program is `oops, something wrong with your C++ program.` The *comma operator* , can be used as a part of nay expression. It strings together several expressions. The value of a comma-separated list of expressions is the value of the right-most expression. The values of the other expressions will be discarded.

```
// brackets.cpp

#include <fstream.h>
#include <iomanip.h>
#include <stdlib.h>

int main()
{
   ifstream source;
   char c;

   unsigned long lcurly = 0, rcurly = 0;
   unsigned long lpar = 0, rpar = 0;
   unsigned long lsquare = 0, rsquare = 0;

   source.open("my.cpp");

   if(!source)
   {
   cerr << "Failed to open my.cpp" << endl;
   }

   source >> resetiosflags(ios::skipws);

   while(source >> c, !source.eof())  // using the comma operator
   {
   if(c == '{') lcurly++;
   if(c == '}') rcurly++;
   if(c == '(') lpar++;
   if(c == ')') rpar++;
   if(c == '[') lsquare++;
   if(c == ']') rsquare++;
   }

   source.close();

   cout << "The number of { is: " << lcurly << endl;
```

```
  cout << "The number of } is: " << rcurly << endl;
  cout << "The number of ( is: " << lpar << endl;
  cout << "The number of ) is: " << rpar << endl;
  cout << "The number of [ is: " << lsquare << endl;
  cout << "The number of ] is: " << rsquare << endl;

  if(lcurly != rcurly)
  {
  cout << "oops, something wrong with your C++ program\n";
  cout << "curly brackets do not match" << endl;
  }

  if(lpar != rpar)
  {
  cout << "oops, something wrong with your C++ program\n";
  cout << "parentheses do not match";
  }

  if(lsquare != rsquare)
  {
  cout << "oops, something wrong with your C++ program\n";
  cout << "rectangular brackets do not match";
  }

  return 0;
}
```

9.3.5 Writing File to Another File

The following program writes a given file named OLD.TXT to another file named NEW.TXT. If the file OLD.TXT does not exist, we obtain the error message:

```
cannot open OLD.TXT for input
```

This program is a C++ version. We recall that all the members (data and functions) of istream and ostream are part of objects of the fstream class.

```
// oldnew.cpp

#include <fstream.h>

int main()
{
   char ch;

   ifstream f1("OLD.TXT");
   ofstream f2("NEW.TXT");

   if(!f1) cerr << "cannot open OLD.TXT for input";
   if(!f2) cerr << "cannot open NEW.TXT for output";

   while((f2 != NULL) && (f1.get(ch) != NULL))   // && logical AND
   f2.put(ch);

   f1.close();
   f2.close();

   return 0;
}
```

The line

```
while((f2 != NULL) && (f1.get(ch) != NULL))
```

can also be written as

```
while(f2 && f1.get(ch))
```

9.4 Binary File Operations

By default, file stream operations are performed in text mode. In some cases we have to perform binary file operations. When a program reads a file in text mode, the program considers the ASCII value 26 (`CTRL-Z`) as the end of the file. If a program tries to read a binary file, it is very likely that the program may encounter this value midway through the file, causing the `eof` function to evaluate as true. To solve this problem, we have to open the file in binary mode. Consider the following program. It writes the integer value to the file `data.dat` using binary file operations.

```
// binary1.cpp

#include <fstream.h>

int main()
{
   int count;
   int num;

   ofstream values("bdata.dat",ios::binary);

   if(values.fail())
   cerr << "Error opening the file bdata.dat" << endl;
   else
   {
   for(count = 1; count < 5; count++)
   {
   num = count*100;
   values.write((char*) &num, sizeof(int));
   }
   values.close();
   }

   return 0;
}
```

When one tries to display the contents of the file `data.dat` using the `type` command, the operation will fail, since the file `data.dat` is a binary and not an ASCII file. The following program shows how to handle this case.

```
// binary2.cpp

#include <fstream.h>
#include <iomanip.h>

int main()
{
   int num;

   ifstream values("bdata.dat",ios::binary);

   while(!values.eof())
   {
   values.read((char*) &num, sizeof(int));
   cout << num << endl;
   }

   values.close();

   return 0;
}
```

9.5 The Functions seekp and seekg

Most programs performs sequential I/O, beginning at the start of the file and working to the end. Random file operations, on the other hand, let the programs perform read and write file operations from any location within the file. The program moves the file pointer to the desired location in the file using the **seekg** and **seekp** member function. The **seekg** function moves the input (get) file pointer, while the **seekp** function moves the output (put) file pointer. Thus the function **seekp**

```
ostream& seekp(streampos)
```

moves to an absolute position. The function **seekg**

```
istream& seekg(streampos)
```

moves to an absolute position. In other words to position file pointers for random input operations we can use the **seekg** member function. Likewise, to position the file pointer for random file output, we can use the **seekp** function. The following small program shows an applications of these two functions.

```
// seekgp.cpp

#include <fstream.h>

int main()
{
   fstream letters("letters.dat",ios::in | ios::out);

   letters.seekp(10,ios::beg);

   for(char letter = 'i'; letter <= 'z'; letter++)
     letters << letter;

   letters.seekg(0, ios::beg);

   while(!letters.eof())
   cout.put((char)letters.get());

   letters.close();
   return 0;
}
```

9.6 The Function isatty

The function

```
int isatty(int handle);
```

checks for device type. It determines whether handle is associated with any one of the following character devices: a terminal, a console, a printer, a serial port

If the device is a character device, isatty returns a nonzero integer. If it is not such a device, isatty returns 0. The following program shows an application of this function.

```
/* myisatty.c */

#include <unistd.h>
#include <stdio.h>
#include <stdlib.h>
#include <termios.h>

void main(void)
{
   FILE* f;

   /* make sure that stdin is a terminal */
   if(!isatty(STDIN_FILENO))
   printf("STDIN is not a terminal.\n");
   else
   printf("STDIN is a terminal.\n");

   /* check for different device */
   f = fopen("/dev/cua1","rw");
   if(f == NULL)
   {
   printf("error opening /dev/cua1\n");
   exit(-1);
   }

   if(!isatty(fileno(f)))
   printf("/dev/cua1 is not a terminal.\n");
   else
   printf("/dev/cua1 is a terminal.\n");

   fclose(f);
}
```

9.7 The Function getenv

The function

```
char* getenv(const char* name);
```

gets a string from environment. It returns the value of a specified variable. On DOS, `name` must be uppercase. On other systems, `name` can be either uppercase or lowercase. `name` must not include the equal sign `=`. If the specified environment variable does not exist, `getenv` returns a `NULL` pointer. On success, `getenv` returns the value associated with `name`.

The following program shows an application of this function.

```
// mygetenv.cpp

#include <stdio.h>
#include <stdlib.h>

int main(void)
{
   char* s1 = NULL;
   s1 = new char [100];
   s1 = getenv("PATH");
   printf("PATH = %s\n",s1);
   delete s1;

   char* s2 = NULL;
   s2 = new char [10];
   s2 = getenv("OS");
   printf("OS = %s\n",s2);
   delete s2;

   char* s3 = NULL;
   s3 = new char [20];
   s3 = getenv("lib");
   printf("lib = %s\n",s3);
   delete s3;

   return 0;
}
```

Chapter 10

Applications in Finance

10.1 Introduction

In this chapter we give a collection of programs with applications to finance. The program `payroll.cpp` calculates the hours worked and the corresponding weekly payment for employees. The program `interest.cpp` asks for the loan, interest rate and monthly payment. The output is the loan balance. The loan granted, the interest rate, the monthly payment and the date of the first payment are inserted into the program `amort1.cpp`. As output we obtain the table of the monthly payment, the principal paid and the last installment. The program `Loan.cpp` illustrates the ease with which the date class can be used just like any built in data type. It considers the case where a person takes out a loan at a fixed interest rate which is compounded daily. The person makes a monthly payment of a certain amount on a fixed day in the month (for example the first day of each month). The program calculates the date on which the last payment will be made and the actual amount required for the last payment. Periodic savings and compound amounts are studied in the program `sav.cpp` and `compp.cpp`. The class concept is applied. We implement the Floyd-Warshall algorithm in the program `floyd.cpp`. The Floyd-Warshall algorithm is used to solve the all-pairs shortest path problem in a network. Then we give an implementation of the Dijkstra algorithm. Finally linear programming and the Simplex method are described and an implementation is given.

10.2 Payroll

A firm has three employees Edward, Joe and Mike. Edward is paid \$ 2.50 an hour, Joe is paid \$ 6.00 an hour and Mike is paid \$ 4.25 an hour. The employees worked the following hours weekly

Edward: 4.0 h, 5.3 h, 10.0 h, 11.0 h
Joe: 1.0 h, 3.0 h
Mike: 3.4 h, 6.7 h, 12.3 h, 3.4h, 6.0 h, 3.0 h, 12.0h

We have to calculate the hours worked and the weekly payment. The output should display the name of the employee, the hours worked weekly and the payment.

The data file `payroll.dat` is given by

```
4.0 5.3 10.0 11.0 2.50 Edward 0
1.0 3.0 6.00 Joe 0
3.4 6.7 12.3 3.4 6.0 3.0 12.0 4.25 Mike
```

The last number before the name is the payrate per hour.

The program is a C program. It does not involve any methods from C++. The function `fopen` opens a stream while the value `r` in the mode string opens for reading only. The function `getc` gets character from stream. The function `fscanf` scans and formats input from a stream. Finally, `printf` writes formatted output to `stdout`. The function `eof` checks for end-of-file.

The output is

```
Edward has worked 30.30 hours. Pay $75.75

Joe has worked 4.00 hours. Pay $24.00

Mike has worked 46.80 hours. Pay $198.90
```

```
// payroll.cpp

#include <stdio.h>

int main()
{
    const int NAME_LENGTH = 15;

    FILE *data;    /* declare a file called data */

    char name[NAME_LENGTH];

    float rate = 0.0,  /* payrate */
          hours = 0.0, /* hours worked per day */
          sum = 0.0;   /* the running total of hours per week */
    data=fopen("payroll.dat","r"); /* open the file payroll.dat */

    do
    {                                /* starts loop */
    hours=0.0; sum=0.0; rate=0.0;    /* initialize variables each */
    while(fscanf(data,"%f",&hours)) /* time while loop is run */
    {                                /* reach each number on the */
    sum = sum + hours;               /* line in the data file once */
    }                                /* it sees the first character */
    /* it will stop reading */
    fscanf(data, "%s\n", &name);     /* read the name */
    sum = sum - hours;               /* sum reads all the numbers, */
    /* to have the correct sum */
    /* the last number read */
    rate = hours;                    /* assign the last number read */
                                     /* to the rate */
    printf("%s has worked %2.2f hours. Pay $%3.2f\n",
    name,sum,sum*rate);
    }
    while(getc(data) != EOF); /* procedure until the end of the file */
    /* this will find the actual EOF */
    /* and terminate the loop */

    return 0;
}
```

10.3 Loan and Interest

In the following program we enter the loan given, the interest rate and the monthly payment. For example, let us assume that a loan of $30000,- is granted. The interest rate is 9.75%. The monthly fixed payment is $360,-. We calculate the interest for one month , the amount repaid and the loan balance after one month. The output is

```
Interest per month = $243.75
Principle repaid = $116.25
Loan balance = $29883.75
```

The `class Loan` contains the data type

```
float intrate, monthp, principal;
```

where `intrate` is the interest rate (9.75), `monthp` is the monthly payment ($ 360,-) and `principal` is the loan of $ 30 000. The

```
class Loan
```

includes four member functions. The member function `getdata` gets the data of the principal, interest rate, and monthly payment. The member function `fintpmnt` calculates the interest per month. The function `fprpaid` calculates the difference bewteen the monthly payment and the interest payment per month. We find the balance using the member function `fbalance`. L is an object of class Loan. The operator . is the member selection operator. Thus `L.getdata` selects the member function `getdata`. The output of the program is the interest for the month, the principal repaid and the loan balance.

```
// interest.cpp

#include <iostream.h>   // for cout, cin

  // Definition of class Loan
class Loan
{
   private:
      double principal, int_rate, monthp;
      double intpmnt,  prpaid,  balance;

   public:
      Loan();              // constructor
      void getdata();      // member function of class Loan
      double fintpmnt(); // member function of class Loan
      double fprpaid();  // member function of class Loan
      double fbalance(); // member function of class Loan
};

// Implementation of class Loan
Loan::Loan() { }

void Loan::getdata()
{
   cout << " Enter loan in $ = ";
   cin >> principal;  cout << "\n";
   cout << " Enter interest rate in % = ";
   cin >> int_rate;  cout << "\n";
   cout << " Enter monthly payment in $ = ";
   cin >> monthp;  cout << "\n";
}

double Loan::fintpmnt()
{
   intpmnt = principal*(int_rate/100.0)/12.0;
   return intpmnt;
}

double Loan::fprpaid()
{
   prpaid = monthp - intpmnt;
   return prpaid;
}

double Loan::fbalance()
```

```
{
   balance = principal - prpaid;
   return balance;
}

int main()
{
   double intpmnt, balance, prpaid;

   Loan L;          // L is object of class Loan
   L.getdata();     // .  member selection operator

   intpmnt = L.fintpmnt();
   prpaid = L.fprpaid();
   balance = L.fbalance();

   cout << " Interest for the month = $" << intpmnt << "\n";
   cout << " Principal repaid = $" << prpaid << "\n";
   cout << " Loan balance = $" << balance << "\n";

   return 0;
}
```

10.4 Amortization

Assume a loan of $5000,- is granted. The interest rate is 10.0 %. The repayment starts in first of January 1995. The monthly payment is $800. Then we want to calculate the repayment for each month and the interest paid each month and the total interest paid.

For example, with the data given above we obtain the output

```
Enter Loan = $5000
Enter Interest Rate = 10.0
Enter Monthly Payment = $800
Enter this Year = 1995

1995 JAN 5000 800 41.668 758.333 4241.66
1995 FEB 4241.667 800 35.347 764.653 3477.014
1995 MAR 3477.013 800 28.975 771.025 2705.989
1995 APR 2705.989 800 22.55 777.45 1928.539
1995 MAY 1928.538 800 16.071 783.929 1144.61
1995 JUN 1144.610 800 9.538 790.462 354.148

1995 interest paid = $154.15 principal paid = $4654.851

Last installment is of amount $354.15
```

The function `setprecison(n)` (n: integer) sets the floating-point precision to `n`.

```
//amort1.cpp

#include <iostream.h>
#include <iomanip.h>   // for setprecision

    // Declaration of class Loan
class Loan
{
   protected:
      float int_rate, mo_pmnt, principal;
   public:
      Loan();
      Loan(float,float,float);
      friend void table();
      float mo_int, mo_bal, adlppl;
      float cum_int, cum_prncpl;
};

Loan::Loan() {  }

Loan::Loan(float mrt,float mpt,float prl)
{
    int_rate = mrt;  mo_pmnt = mpt;  principal = prl;
}

float pr, mr, mp, tointpmnt;
int yr;

void init()
{
   cout << "Enter Loan = $"; cin >> pr;
   cout << "Enter Interest Rate = "; cin >> mr;
   cout << "Enter Monthly Payment = $"; cin >> mp;
   cout << "Enter this Year = "; cin >> yr;
};

void table()
{
    char* q = "JANFEBMARAPRMAYJUNJULAUGSEPOCTNOVDEC";
    char* p;
    float x, y, z, t, u, rate, balance, w;
    init();
    Loan L(mr,mp,pr);

    p = q;
```

```
    Loan yd;  Loan md;
    md.adlppl = 0.0;
    w = md.adlppl;
    yd.cum_int = 0.0;
    yd.cum_prncpl = 0.0;
    rate = L.int_rate/1200.0;
    balance = L.principal;

    while(balance >= L.mo_pmnt) {
    while((*p !='\0') && (balance >= L.mo_pmnt)) {
    u = balance;
    md.mo_int = u*rate;
    z = md.mo_int;
    x = L.mo_pmnt-md.mo_int;
    t = x+w;
    md.mo_bal = u-t;
    y = md.mo_bal;

    cout << yr <<" "<< *p << *(p+1) << *(p+2) << " ";
    cout << setprecision(3) << u << " " << t+z << " " << z
        << " " << t << " " << y << "\n";

    balance = y;
    yd.cum_int += md.mo_int;
    yd.cum_prncpl += x+w;
    p += 3;
    }
    cout << "\n\n";
    cout << setprecision(2) << yr << " interest paid = $"
     << yd.cum_int;
    cout << setprecision(3) << " principal paid = $"
     << yd.cum_prncpl;
    cout << "\n\n";
    yr +=1;
    p = q;
    }
    cout << "\n Last installment is of amount $";
    cout << setprecision(2) << balance << "\n";
};

void main()
{
    table();
}
```

10.5 Calculating the Loan Period

Calculates the date and amount of the last payment of a loan taken out at a fixed interest rate which is compounded daily and payed off by a fixed monthly payment on a specific day of each month. The program includes "`date.cpp`" (see chapter 8).

```
// LOAN.CPP

#include <iostream.h>
#include <iomanip.h>
#include "date.cpp"

int main()
{
  float debt, rate, monthlypayment;
  cout << "Enter initial sum owed: R";
  cin  >> debt;
  cout << "Enter monthly payment: R";
  cin  >> monthlypayment;
  cout << "Enter anual interest rate (%): ";
  cin  >> rate;

  Date date0;
  cout << "Enter date loan taken (e.g. 23 2 1994): ";
  cin  >> date0;

  int payday;
  cout << "Enter day on which monthly installments are payed " << endl
       << "(eg. 1 for 1st of each month): ";
  cin  >> payday;

  Date currentdate(date0);

  float lastpayment=0;

  while(debt > 0)
  {
    int day, year;
    Month month;
    ++currentdate;

    debt += debt*rate/(100*currentdate.DaysInYear());
```

```
      currentdate.GetDate(day,month,year);

      if(day==payday)
      {
      if(debt > monthlypayment)
      debt -= monthlypayment;
      else
      {
      lastpayment = debt;
      debt = 0;
      }
      }
   }   // end while

   cout << "The last payment of R" << setprecision(2)
        << lastpayment << " will be made on "
        << currentdate << endl;

   return 0;
}
```

Take note how the user uses the date class as if it were a built-in data type, without being confronted with any of the implementation details of the class.

10.6 Periodic Savings

The following program calculates quantities of interest for a schedule of n periodic payments p into a savings account with annual interest i compounded at the time of each payment. The other variables of importance are the future value S_n after n payments periods. With

$$\lambda := \frac{im}{1200}$$

where m is the length of a payment period in month the following calculations can be made:

(1) For given p, i and n, the quantity S_n is calculated

$$S_n = \frac{(1+\lambda)^{n+1} - (1+\lambda)}{\lambda} p$$

(2) For given i, n and S_n, the quantity p is calculated

$$p = \frac{\lambda}{(1+\lambda)^{n+1} - (1+\lambda)} S_n$$

(3) For given i, p and S_n, the quantity n is calculated

$$n = \frac{\ln(\frac{\lambda S_n}{p} + (1+\lambda))}{\ln(1+\lambda)} - 1$$

For example, let $p = 500$, $i = 10.0\%$, $n = 5$, and $m = 2$, then

$$S_n = 2627.813$$

```
// sav.cpp
// periodic savings

#include <iostream.h>
#include <ctype.h>    // for toupper
#include <math.h>     // for pow, log
#include <conio.h>    // for getch
#include <iomanip.h>

class Savings
{
   private:
      double a, b, c, d;
   public:
      Savings(double,double,double,double);
      double calSn();
      double calp();
      double caln();
};

Savings::Savings(double u,double v,double w,double x)
{
   a = u;  b = v;  c = w;  d = x;
}

double Savings::calSn()
{
   double lam = b*d/1200.0;
   return a*(pow((1.0 + lam),c+1) - (1.0 + lam))/lam;
}

double Savings::calp()
{
   double lam = a*d/1200.0;
   return c*lam/(pow((1.0 + lam),b+1) - (1.0 + lam));
}

double Savings::caln()
{
   double lam = b*d/1200.0;
   return (log(lam*c/a + (1.0 + lam)))/(log(1 + lam)) - 1.0;
}

int main()
{
```

```
double u, v, w, x;
char cmd;
do
{
cout << endl << endl;
cout << "Chart desired: ";
cout << "Future_value Payment_amount Number_of_payments Quit ";
cout << endl;
cout << "Press first letter of the chart you want: ";
cmd = toupper(getch());
cout << endl;

switch(cmd)
{
case 'F':
cout << "enter the periodic payment: ";
cin >> u;
cout << "enter the interest rate in %: ";
cin >> v;
cout << "enter the number of payments: ";
cin >> w;
cout << "enter the length of a payment period in month: ";
cin >> x;
Savings SN(u,v,w,x);
cout << "Future value: ";
cout << setprecision(2) << SN.calSn();
break;

case 'P':
cout << "enter the interest rate in % : ";
cin >> u;
cout << "enter the number of payments: ";
cin >> v;
cout << "enter future value: ";
cin >> w;
cout << "enter the length of a payment period in month: ";
cin >> x;
Savings Sp(u,v,w,x);
cout << "periodic payments: ";
cout << Sp.calp();
break;

case 'N':
cout << "enter the periodic payment: ";
cin >> u;
```

```
    cout << "enter the interest rate in %: ";
    cin >> v;
    cout << "enter the future value: ";
    cin >> w;
    cout << "enter the length of a payment period in month: ";
    cin >> x;
    Savings Sn(u,v,w,x);
    cout << "number of payments: ";
    cout << Sn.caln();
    break;

    case 'Q': break;

    default : cout << "invalid choice. try again " << endl;
    }
    } while (cmd != 'Q');

    return 0;
}
```

10.7 Compound Amount

The following program applies to an amount of principal, S_0, that has been placed in an account and compounded periodically at an annual interest rate i with no further deposits. Other important variables in the problem are the future value S_n after n compounding periods and the amount of interest accrued I. The interest rate is contained in the quantity

$$\lambda := \frac{im}{1200}$$

where m is the compounding period in months. Each of the variables can be calculated from the others as follows:

(1) For given S_0, i, m and n, the quantity S_n is calculated as

$$S_n = S_0(1+\lambda)^n \tag{1}$$

(2) For given S_n, i, m and n, the quantity S_0 is calculated as

$$S_0 = S_n(1+\lambda)^{-n} \tag{2}$$

(3) For given S_0, S_n, m, and n, the quantity i is calculated as

$$\lambda = \left(\frac{S_n}{S_0}\right)^{1/n} - 1 \tag{3}$$

where $i = 1200\lambda/m$.

(4) For given S_0, S_n and i, the quantity n is calculated as

$$n = \frac{\ln\left(\frac{S_n}{S_0}\right)}{\ln(1+\lambda)}\,. \tag{4}$$

(5) For given S_0, i and n, the quantity I is calculated as

$$I = S_0((1+\lambda)^n - 1)\,. \tag{5}$$

We implement equations (1), (2) and (4).

```
// comp.cpp
// compound amount

#include <iostream.h>
#include <ctype.h>    // for toupper
#include <math.h>     // for pow, log
#include <conio.h>    // for getch

class Compound
{
   private:
      double a, b, c, d;
   public:
      Compound(double,double,double,double);
      double calSn();
      double calS0();
      double caln();
};

Compound::Compound(double u,double v,double w,double x)
{
   a = u;  b = v;  c = w;  d = x;
}

double Compound::calSn()
{
   double lam = c*b/1200.0;
   return  a*pow((1.0 + lam),d);
}

double Compound::calS0()
{
   double lam = a*b/1200.0;
   return c*pow((1.0 + lam),-d);
}

double Compound::caln()
{
   double lam = b*d/1200.0;
   return (log(c/a))/(log(1.0 + lam));
}

int main()
{
   double u, v, w, x;
```

```
char cmd;

do
{
cout << endl << endl;
cout << "Chart desired: ";
cout << "Future_value ";
cout << "Principal_value ";
cout << "Number_of_compounding_periods ";
cout << " Quit ";
cout << endl;
cout << "Press first letter of the chart you want: ";
cmd = toupper(getch());
cout << endl;

switch(cmd)
{
case 'F':
cout << "amount of principal: ";
cin >> u;
cout << "length of compounding period in months: ";
cin >> v;
cout << "enter the interest rate: ";
cin >> w;
cout << "number of compounding periods: ";
cin >> x;
Compound SN(u,v,w,x);
cout << "Future value: ";
cout << SN.calSn();
break;

case 'P':
cout << "enter the interest rate in % : ";
cin >> u;
cout << "length of compouding periods in months: ";
cin >> v;
cout << "enter future value: ";
cin >> w;
cout << "number of compounding periods: ";
cin >> x;
Compound SO(u,v,w,x);
cout << "amount of principal: ";
cout << SO.calSO();
break;
```

```
    case 'N':
    cout << "enter principal value: ";
    cin >> u;
    cout << "enter the interest rate in %: ";
    cin >> v;
    cout << "enter future value: ";
    cin >> w;
    cout << "enter the length of a compounding period in months: ";
    cin >> x;

    Compound n(u,v,w,x);
    cout << "number of payments: ";
    cout << n.caln();
    break;

    case 'Q': break;

    default : cout << "invalid choice. try again " << endl;
    }
    } while(cmd != 'Q');

    return 0;
}
```

10.8 The Floyd–Warshall Algorithm

Given a network, the all-to-all shortest path problem is to find the total cost of going from every node to every other node in the network. For some pairs of nodes, this can be found by inspecting the network. In a large network that may consist of hundreds of arcs so that finding the shortest path between various nodes may be complicated. A method to solve this problem is the *Floyd–Warshall algorithm.* It uses a two-dimensional array to repreent the cost of travelling from every node to every other node. Each row of the two-dimensional array corresponds to an originating node, and each column corresponds to a destination node.

The Floyd–Warshall algorithm begins by initializing the entries in the cost array with the costs of the arcs in the network. Let n be the number of nodes. The initialization is as follows:
1) Initialize the diagonal elements of the two dimensional array a to zero, i.e.

$$a_{00} = a_{11} = \ldots a_{n-1,n-1} = 0$$

This reflects the fact that travel from one node to itself is costless.
2) Each arc in the network connects some node i and some node j. For each arc, initialize the value c_{ij} in the cost matrix C to the cost of that arc.
3) All other entries in the array should be set to some large value that exceeds the total cost of any final path in the network (for example we set it to be 20000).

The Floyd–Warshall algorithm then examines every pair of nodes ij $(i \neq j)$, and for each node pair, tests whether the current cost of going from i to j can be reduced by using node 0 as an intermediate node. This corresponds to finding the following minimum

$$\min(c_{ij}, c_{i0} + c_{0j})$$

Once this is done for all ij pairs, the updated matrix C now holds the cost of the shortest paths that use only node 0 as a possible intermediate node.

As an example consider the matrix

$$C = \begin{pmatrix} 0 & 5 & 3 & 20000 & 1 \\ 2 & 0 & 20000 & 2 & 1 \\ 6 & 20000 & 0 & 2 & 20000 \\ 20000 & 1 & 1 & 0 & 8 \\ 1 & 3 & 20000 & 1 & 0 \end{pmatrix}$$

The resulting matrix is given by

$$C_F = \begin{pmatrix} 0 & 3 & 3 & 2 & 1 \\ 2 & 0 & 3 & 2 & 1 \\ 5 & 3 & 0 & 2 & 3 \\ 3 & 1 & 1 & 0 & 2 \\ 1 & 2 & 2 & 1 & 0 \end{pmatrix}$$

```
// floyd.cpp

#include <iostream.h>
#include <stdio.h>
#include <limits.h>

void floyd_warshall(int**,int);
int minint(int,int);
void show_Cost(int**,int);

int main()
{
   int i, j, nodes;

   int **Cost;

   cout << "number of nodes: ";
   cin >> nodes;

   Cost = new int *[nodes];
   for(i=0;i<nodes; i++)
   {
   Cost[i] = new int [nodes];
   }

   for(i=0; i<nodes; i++)
   {
   for(j=0; j<nodes; j++)
   Cost[i][j] = 0;
   }

   for(i=0; i<nodes; i++)
   for(j=0; j<nodes; j++)
   if(i != j)
   {
   cout << "Cost[" << i << "][" << j << "] = ";
   cin >> Cost[i][j];
   }

   floyd_warshall(Cost,nodes);
   show_Cost(Cost,nodes);

   return 0;
}
```

```
void floyd_warshall(int **m,int nodes)
{
   int i, j, k;
   for(k=0; k<nodes; k++)
   for(i=0; i<nodes; i++)
   for(j=0; j<nodes; j++)
   if(i != j)
   m[i][j] = minint(m[i][j], m[i][k]+m[k][j]);
}

int minint(int x,int y)
{
   if(x < y) return x;
   return y;
}

void show_Cost(int **m,int nodes)
{
   int i, j;
   cout << "Cost matrix is: " << endl;

   for(i=0; i<nodes; i++)
   {
   for(j=0; j<nodes; j++)
   cout << m[i][j] << " ";
   cout << endl;
   }
}
```

10.9 The Dijkstra Algorithm

A frequent application of graphs is to find the shortest path between vertices in a weighted directed graph. The graph might represent a circuit, with edge weights being lengths of connectors in microns or inches; it might represent the phases of a construction project, with the edge weights being time in days. With the understanding that the edge weights are generalized "costs" in the application, the problem is often described as that of finding the minimum-cost path between one vertex in a graph and all others. We implement the *Dijkstra algorithm*. We use an adjacency matrix representation, in which the edge lengths are the costs (distances, times, or whatever) associated with the edges. We initialize this matrix, called `C` in our program, to very large values, indicating impossibly large costs where there are no edges. We then initialize an array called `dist` to equal the first row of the edge matrix. We now seek the shortest parth from the first vertex to all others. Our approach is to identify a set of vertices for which we know the shortest possible distances from vertex 1; we call this "special" set `S`. The first vertex is the default first member of `S`; the shortest possible distance from vertex 1 to vertex 1 is, of course, zero. Now we go through a two-step process, adding vertices to the set `S` until they are all in it.

```
// Dijk2.cpp

#include <iostream.h>

unsigned find(unsigned* S,unsigned n,unsigned x)
{
   int k;
   for(k=0; k<n; k++)
   {
   if(x==S[k]) return 1;
   }
   return 0;
}

int main()
{

   // number of vertices = size of cost matrix
   unsigned n = 6;

  unsigned i, j, w;
  unsigned mindist;
  unsigned* S = NULL;
  S = new unsigned [n];
```

```
// cost matrix
unsigned** C = NULL;
C = new unsigned* [n];
for(i=0; i<n; i++)
C[i] = new unsigned [n];

unsigned* dist = NULL;
dist = new unsigned [n];
unsigned* path = NULL;
path = new unsigned [n];

for(i=0; i<n; i++)  {  S[i] = 0;  }

unsigned inf = 60000;

C[0][0] = 0; C[0][1] = 30; C[0][2] = inf;
C[0][3] = 50; C[0][4] = 40; C[0][5] = 100;
C[1][0] = inf; C[1][1] = 0; C[1][2] = 40;
C[1][3] = inf; C[1][4] = inf; C[1][5] = inf;
C[2][0] = inf; C[2][1] = inf; C[2][2] = 0;
C[2][3] = inf; C[2][4] = 10; C[2][5] = 30;
C[3][0] = inf; C[3][1] = inf; C[3][2] = 10;
C[3][3] = 0; C[3][4] = inf; C[3][5] = inf;
C[4][0] = inf; C[4][1] = inf; C[4][2] = inf;
C[4][3] = 20; C[4][4] = 0; C[4][5] = 70;
C[5][0] = inf; C[5][1] = inf; C[5][2] = inf;
C[5][3] = inf; C[5][4] = inf; C[5][5] = 0;

for(j=0; j<n; j++)
{
dist[j] = C[0][j];
if(C[0][j] == inf)
{ path[j] = 0; }
else
path[j] = 1;
}

for(i=0; i<n-1; i++)
{
mindist = inf;
for(j=1; j<n; j++)
{
if((find(S,n,j) == 0) && (dist[j] < mindist))
{
```

```
  mindist = dist[j];
  w = j;
  }
  }
  S[i] = w;

  for(j=1; j<n; j++)
  {
  if((find(S,n,j) == 0) && (dist[w] + C[w][j] < dist[j]))
  {
  dist[j] = dist[w] + C[w][j];
  path[j] = w;
  }
  }
  }

  // output
  for(i=0; i<n; i++)
  cout << "dist[" << i << "] = " << dist[i] << endl;

  for(i=0; i<n; i++)
  cout << "path[" << i << "] = " << path[i] << endl;

  for(i=0; i<n; i++)
  {
  cout << "S[" << i << "] = " << S[i] << endl;
  }
  return 0;
}
```

The output is

```
dist[0] = 0  dist[1] = 30 dist[2] = 60
dist[3] = 50 dist[4] = 40 dist[5] = 90

path[0] = 1 path[1] = 1 path[2] = 3
path[3] = 1 path[4] = 1 path[5] = 2

S[0] = 1 S[1] = 4 S[2] = 3
S[3] = 2 S[4] = 5 S[5] = 0
```

10.10 Linear Programming and Simplex Method

Linear programming is concerned with the following problem. For N independent variables $x_1, \ldots, x_N$, maximize the function

$$p = a_{01}x_1 + a_{02}x_2 + \cdots + a_{0N}x_N \tag{1}$$

subject to the primary constraints

$$x_1 \geq 0, \qquad x_2 \geq 0, \quad \cdots \quad , x_N \geq 0 \tag{2}$$

and simultaneously subject to $M = m_1 + m_2 + m_3$ additional constraints, m_1 of them of the form

$$a_{i1}x_1 + a_{i2}x_2 + \cdots + a_{iN}x_n \leq b_i \quad (b_i \geq 0)\ ; \quad i = 1, \ldots, m_1 \tag{3}$$

m_2 of them of the form

$$a_{j1}x_1 + a_{j2}x_2 + \cdots + a_{jN}x_n \geq b_j \geq 0\ ; \quad j = m_1 + 1, \ldots, m_1 + m_2 \tag{4}$$

and m_3 of them of the form

$$a_{k1}x_1 + a_{k2}x_2 + \cdots + a_{kN}x_N = b_k \geq 0 \tag{5}$$

where

$$k = m_1 + m_2 + 1, \quad \ldots \quad , m_1 + m_2 + m_3\,.$$

The various a_{ij}'s can have either sign, or be zero. The fact that the b's must all be nonnegative (as indicated by the final inequality in the above three equations) is a matter of convention only, since one can multiply any contrary inequality by -1. There is no particular significance in the number of constraints M being less than, equal to, or greater than the number of unknowns N. Furthermore if p in (1) is to be minimized, it is equivalent to $-p$ being maximized.

The problem is solved in two stages.

(a) Inequalities of the form (3) or (4) are transformed to equalities by adding [in the case of(3)] or subtracting so-called [in the case of(4)] so-called *slack variables* of which nonnegativity obviously are required. We will the denote slack by y_i; $i = 1, 2, \ldots, m_1 + m_2$. Once they are introduced, they are treated on an equal footing with the original variables x_i; at the end of the calculation they are simply ignored.

Next, M artificial variables, denoted by p_i, are introduced. The problem is then put into restricted normal form by rewriting the constraints (3) through (5) in the form

$$p_i = b_i - (a_{i1}x_1 + a_{i2}x_2 + \cdots + a_{iN}x_n + y_i), \qquad i = 1, \ldots, m_1 \tag{6}$$

$$p_j = b_j - (a_{j1}x_1 + a_{j2}x_2 + \cdots + a_{jN}x_n - y_j), \qquad j = m_1 + 1, \ldots, m_1 + m_2 \quad (7)$$

$$p_k = b_k - (a_{k1}x_1 + a_{k2}x_2 + \cdots + a_{kN}x_N) \quad (8)$$

where

$$k = m_1 + m_2 + 1, \ldots, m_1 + m_2 + m_3 \quad (8)$$

The objective function (1) is replaced by a so-called auxiliary objective function

$$p' := -\sum_{i=1}^{M} p_i \,. \quad (9)$$

(b) The problem in the restricted normal form is solved using the *simplex method.* The algorithm can be summarized as follows:

A so-called *tableau* is set up which summarizes (1) and (6) to (9) in the following way:

(i) For the first tableau the y_i in (6) and (7) are denoted as x_{i+n} and the x_j; $j = 1, 2, \ldots, n+m_1+m_2$ are then noted as *right-hand variables.* The p_i in (6) to (8) are denoted as x_j; $j = n + m_1 + m_2 + 1, n + m_1 + m_2 + 2, \ldots, n + 2(m_1 + m_2)$ and noted as *left-hand variables.*

(ii) The first row contains 0 and the a_{0i} in (1) as entries.

(iii) The next $m_1 + m_2 + m_3$ rows each contains the b_i and a_{ij} of one of the equations (3) through (4) as entries.

(iv) The last row's first element is the constant term in (9); the rest of the entries are the coefficients of the x_i (in order of ascending i) in (9).

A left-hand variable is now exchanged with a right-hand variable by performing linear operations on the equations as contained in the tableau; this is done in such a way that the value of p' is reduced. This process is repeated until p' can not be further reduced, i.e. until all the p_i are right-hand variables. (If this situation is not attainable, it signals that there is no feasible basic vector, i.e that the constraints are inconsistent, and the process is terminated.)

Once (9) is maximized, the p_i are removed from the tableau, and the same procedure is repeated in order to maximize (1).

Listed below is the routine. The construction of the routine is as follows:

1. In `main` the input values of M, N_1, m_1, m_2 and m_3 are entered, where N_1 is the number of basic variables [N in (1)] and the other symbols have the same meaning as in (3) to (5). For the rest, `main` only calls some other functions. Throughout the routine `N` represents the total number of basic and slack variables.

2. The variable declared as `double *x` in `display` contains all the variables in the following way:
   The basic variables as `x[i]`, `i`$= 0, 1, \ldots, N_1 - 1$
   The slack variables as `x[i]`, `i`$= N_1, N_1 + 1, \ldots, N - 1$
   The artificial variables as `x[i]`, `i`$= N, N + 1, \ldots, N + M - 1$

3. The coefficients in the objective function as well as those in the constraints are input by `coefficients`. The initial tableau is then set up by `tableau`. The elements of the tableau are entered in the array `a[i][j]`, `i` $= 0, 1, \ldots, M + 1$; `j` $= 0, 1, ..., N$. The elements are ordered in the following way:

   (i) The first column, `a[i][0]`, `i` $= 0, 1, \ldots, M$, contains the constant terms of the objective function (as `a[0][0]`) and the auxiliary objective function (as `a[N][0]`) as well as the constant terms on the righthand side in the equations (6) to (8) for the artificial variables.

   (ii) In the first row, the coefficients of the x_j in the objective function (1) are entered as `a[0][j]`, `j` $= 1, 2, \ldots, N$.

   (iv) In the `i`-th row, `i` $= 1, 2, \ldots, M$, the coefficients of the x_j on the right-hand sides of the equations (6) to (8) are entered as `a[i][j]`, `j` $= 1, 2, \ldots, N$.

   (v) In the $M+1$-st row, the coefficients of the x_j on the righthand side of the auxiliary objective function (9) are entered as `a[M][j]`, `j` $= 1, 2, \ldots, N$.

   The left- and right-hand variables, numbered as in 2, are respectively recorded by `tableau` in the arrays `lhv` and `rhv`.

4. The function `showtab` displays the initial tableau as well as the tableau after each interchange of variables. The use of this function is optional.

5. The simplex algorithm is performed by the function `simplex`. It first maximizes the auxiliary objective function and then the given objective function. At each step of the algorithm, `pivot` is used to find the pivot element and `exchange` is used to exchange the requisite left- and right-hand variables. Once the auxiliary objective function is maximized, `test` is used to establish whether a feasible starting solution has been obtained. Before the given objective function is maximized, `remove` is used to remove artificial variables as right-hand variables. `simplex` returns the variable `flag` as well as the final tableau `a` to `main`. The value of `flag` indicates whether a solution to the given problem exists or not.

6. The final result is displayed by `display`.

```
// simplex.cpp
// Linear Programming with the simplex method

#include <iostream.h>
#include <math.h>
#include <iomanip.h>
#include <conio.h>

const double eps = 1.0e-10;  // Absolute precision
const double G = 1.0e100;    // Very large number

void coefficients(double**,int,int,int,int,int);
void tableau(double**,int*,int*,int,int,int,int,int);
void showtab(double**,int*,int*,int,int,int,int);
void simplex(double**,int*,int*,int,int,int&,int);
void pivot(int,double**,int,int,int&,int&,int&);
void exchange(double**,int*,int*,int,int,int,int);
void test(double**,int*,int*,int,int,int&);
void remove(double**,int*,int,int);
void display(double**,int*,int,int,int,int);

void main()
{
   int N, N1, m1, m2, m3, M, parity, tab;
   cout << endl
        << "Enter  1 if objective function must be MAXIMIZED" << endl
        << "Enter -1 if objective function must be MINIMIZED" << endl;
   cin >> parity;
   cout << endl
        << "Enter 0 if only final results must be displayed" << endl
        << "Enter 1 if intermediate tableaux must be displayed"
        << endl;
   cin >> tab;
   cout << endl << "Number of variables? ";
   cin >> N1;
   cout << endl << "Number of constraints of type <= : ";
   cin >> m1;
   cout << endl << "Number of constraints of type >= : ";
   cin >> m2;
   cout << endl << "Number of constraints of type = : ";
   cin >> m3;
   M = m1 + m2 + m3;  N = N1 + m1 + m2;

   double **a = NULL;           // Elements of tableau
   a = new double*[N + 1];
```

```
   int row_nr;
   for(row_nr = 0; row_nr < M + 2; row_nr++)
   a[row_nr] = new double[N + 1];
   int *lhv = new int[M];       // Left-hand variables
   int *rhv = new int[N];       // Right-hand variables

   coefficients(a,m1,m2,N1,M,parity);    // Coefficients of objective
                                         // function and constraint
                                         // equations are read in
   tableau(a,lhv,rhv,m1,m2,N1,N,M);      // Set up initial tableau
   int flag = 0;
   if(tab == 1)
   {
   cout << endl << "Initial tableau:" << endl;
   showtab(a,lhv,rhv,N,M,0,flag); // Display initial tableau
   }
   simplex(a,lhv,rhv,N,M,flag,tab);  // Simplex algorithm
   display(a,lhv,M,N1,flag,parity);  // Display results
}

// Coefficients in objective function and constraints are entered
void coefficients(double** a,int m1,int m2,int N1,int M,int p)
{
   cout << endl
        << "READ IN COEFFICIENTS IN OBJECTIVE FUNCTION " << endl;
   cout << endl << "Constant term in objective function = ";
   cin >> a[0][0];
   a[0][0] *= p;
   int i;
   for(i=1; i < N1+1; i++)
   {
   cout << endl
        << "Coefficient of x[" << i << "] in objective function = ";
   cin >> a[0][i];
   a[0][i] *= p;
   }
   cout << endl;

   if(m1 != 0)
   {
     cout << endl << endl
          << "READ IN COEFFICIENTS OF CONSTRAINTS OF TYPE <= " << endl
     for(i=1; i < m1 + 1; i++)
     {
     int j;
```

```
    for(j=1; j < N1+1; j++)
    {
    cout << endl << "Coefficient of x[" << j << "] in "
         << i << "-th constraint equation = ";
    cin >> a[i][j];
    }
    cout << endl << "Right-hand side of "
         << i << "-th constraint equation = ";
    cin >> a[i][0];
    cout << endl;
    }
  }

  if(m2 != 0)
  {
    cout << endl << endl
         << "READ IN COEFFICIENTS OF CONSTRAINTS OF TYPE >= " << endl;
    for(i=m1+1; i < m1+ m2+1; i++)
    {
    int j;
    for(j=1; j < N1+1; j++)
    {
    cout << endl << "Coefficient of x[" << j << "] in "
         << i << "-th constraint equation = ";
    cin >> a[i][j];
    }
    cout << endl << "Right-hand side of "
         << i << "-th constraint equation = ";
    cin >> a[i][0];
    cout << endl;
    }
  }

  if(M- m1-m2 != 0)
  {
  cout << endl << endl
       << "READ IN COEFFICIENTS OF CONSTRAINTS OF TYPE = " << endl;
  for(i=m1+m2+1;  i < M+1;  i++)
  {
     for(int j=1; j < N1+1; j++)
     {
     cout << endl << "Coefficient of x[" << j << "] in "
          << i << "-th constraint equation = ";
     cin >> a[i][j];
     }
```

```
        cout << endl << "Right-hand side of "
             << i << "-th constraint equation = ";
        cin >> a[i][0];
        cout << endl;
        }
    }
}

// Initial tableau is set up
void tableau(double** a,int* lhv,int* rhv,int m1,int m2,int N1,int N,
             int M)
{
   // a[i][0] are right-hand side constants
   // a[0][i] are coefficients of objective function
   // a[M + 1][i] are coefficients of auxilliary objective function
   // Coefficients of slack variables in objective function
   int i;
   for(i=N1+1; i < N+1; i++)
      a[0][i] = 0.0;

   // Coefficients of slack variables in constraints
   for(i=1; i < m1 + 1; i++)
   {
   int j;
   for(j=N1 + 1; j < N + 1; j++)
   {
   a[i][j] = 0.0;
   if(j == i + N1) a[i][j] = 1.0;
   }
   }
   for(i=m1 + 1; i < m1 + m2 + 1; i++)
   {
      for(int j=N1 + 1; j < N + 1; j++)
      {
      a[i][j] = 0.0;
      if(j == i + N1) a[i][j] = - 1.0;
      }
   }
   for(i=m1 + m2 + 1; i < M + 1; i++)
   {
   for(int j = N1 + 1;  j < N + 1;  j++)
   a[i][j] = 0.0;
   }

   // Setting up LP tableau
```

```
      for(i=1; i < M + 1; i++)
      {
      if(a[i][0] >= 0.0)   // Check if rhs constants are positive
      {
      for(int j=1; j < N + 1; j++)
      a[i][j] *= -1.0;     // Make the x[i] right-hand variables
      }
      else
      a[i][0] *= -1.0;
      }
      int j;
      for(j=0; j < N + 1; j++)  // Calculate coefficients in
      {                         // auxilliary objective function
      long double sum = 0.0;
      for(int i=1; i < M + 1; i++)
      sum += a[i][j];
      a[M+1][j] = - sum;
   }
   int k;
   for(k=0; k < M; k++)       // Artificial variables p[i] are
   lhv[k] = k + N + 1;        // recorded as left-hand variables
   for(j=0;  j < N;  j++)     // Variables and slack variables
   rhv[j] = j + 1;            // are recorded as right-hand variables
}

// Current tableau is displayed
void showtab(double** a,int* lhv,int* rhv,int N,int M,int phase,
             int flag)
{
   cout << setprecision(2);           // Format of output is set
   cout << setiosflags(ios::left);
   int i;
   for(i=0; i < M+2-phase; i++)
   {
   cout << endl;
   for(int j = 0;  j < N + 1 - M*phase;  j++)
   cout << setw(5) << a[i][j] << '\t';
   }
   cout << endl << endl << "left-hand variables:   ";
   for(i = 0; i < M;  i++)
   cout << lhv[i] << " ";
   cout << endl << "right-hand variables:  ";
   for(i = 0;  i < N - M*phase;  i++)
   cout << rhv[i] << " ";
   cout << endl << "flag = " << flag;
```

```
   int cont;
   cout << endl << endl << "Enter any number to continue ";
   cin >> cont;
}

// Simplex algorithm is performed
void simplex(double** a,int* lhv,int* rhv,int N,int M,int& flag,int tab)
{
  flag = 0;
  // Phase one : maximizing auxiliary objective funtion
  while(flag == 0)
  {
    int pivot_col, pivot_row;
    pivot(M+1,a,N,M,pivot_col,pivot_row,flag); // Find pivot element
    if(flag == 0)
    {
    exchange(a,lhv,rhv,pivot_col,pivot_row,M,N);// Exchange left-hand
                                               // variable with
                                               //right-hand one
    if(tab == 1) showtab(a,lhv,rhv,N,M,0,flag);
    cout << endl << endl;
    }
  }

  if(flag == 1)                        // Test if feasible starting
  {                                    // solution has been found
  flag = 0;
  test(a,lhv,rhv,M,N,flag);
  }

  // Phase two : maximizing given objective funtion
  if(flag == 0)
  {
  remove(a,rhv,N,M);  // Remove artificial variables
  if(tab == 1) showtab(a,lhv,rhv,N,M,1,flag);
  while(flag == 0)
  {
  int pivot_col, pivot_row;
  // Find pivot element
  pivot(0,a,N-M,M,pivot_col,pivot_row,flag);
  if(flag == 0)
  {
   // Exchange left-hand variable with right-hand one
  exchange(a,lhv,rhv,pivot_col,pivot_row,M,N);
  if(tab ==1) showtab(a,lhv,rhv,N,M,1,flag);
```

```
  cout << endl << endl;
  }
  }
  if(flag == 1) flag = 0;
  }
}

// Pivot element is established
void pivot(int M_ob,double** a,int N,int M,int& pc,int& pr,int& flag)
{
   pc = 0;
   double maxp = 0.0;
   flag = 1;
   int j;
   for(j=1; j < N + 1; j++)
   {
   double pac = a[M_ob][j];
   if(pac > eps)
   {
   flag = 0;
   if(pac > maxp)
   {
   pc = j;
   maxp = pac;
   }
   }
   }
   if(flag == 0)
   {
   flag = 2;
   double mina = G;
   pr = 0;
   for(int i = 1;  i < M + 1;  i++)
   {
   double ac = a[i][pc];
   if(ac < 0.0)
   {
   flag = 0;
   double ratio = fabs(a[i][0]/ac);
   if(ratio < mina)
   {
   pr = i;
   mina = ratio;
   }
   }
```

```
   }
   }
}

// A left-hand variable is exchanged with a right-hand one
void exchange(double** a,int* lhv,int* rhv,int pc,int pr,int M,int N)
{
   // Save pivot element
   double piv_el = a[pr][pc];
   // Save pivot column
   double* piv_col = new double[M + 1];
   int k;
   for(k=0;  k < M + 2;  k++)
   piv_col[k] = a[k][pc];
   // Do exchange
   int i;
   for(i=0; i < M+2; i++) //Replace rows
   {
      if(i != pr)
      {
      double ratio = a[i][pc]/piv_el;
      for(int j=0; j < N+1; j++)
      a[i][j] -= ratio * a[pr][j];
      }
   }
   int j;
   for(j=0; j < N+1; j++)  // Divide pivot element by
   a[pr][j] /= -piv_el;       // negative of pivot
   a[pr][pc] = 1.0/piv_el;
   for(i=0; i < M + 2; i++) // Replace pivot column
   {
      if(i == pr)
      a[i][pc] = 1.0/piv_el;
      else
      a[i][pc] = piv_col[i]/piv_el;
   }

   // Clean up small numbers
   for(i=0; i < M + 2; i++)
     for(j=0; j < N+1; j++)
        if(fabs(a[i][j]) < eps)  a[i][j] = 0.0;

   // Record exchange of rh and lh variables
   int dummy = lhv[pr-1]; lhv[pr-1] = rhv[pc-1]; rhv[pc-1] = dummy;
   delete [] piv_col;
```

```
}

// Test for feasible starting solution
void test(double** a,int* lhv,int* rhv,int M,int N,int& flag)
{
   int i;
   for(i=0; i < M; i++)
   {
   // Exchange left-hand artificial variables for right-hand variables
   if(lhv[i] > N)
   {
   if(a[i+1][0] != 0.0) flag = 1;
   else
   {
   int column = 1;
   while(a[0][column] != 0.0)
   {
   int dummy;
   if(rhv[column - 1] < N)
   {
   exchange(a,lhv,rhv,column,i+1,M,N);
   break;
   }
   column++;
   if(column > N)
   {
   flag = 1;
   break;
   }
   }
   }
   }
   }
}

// Artificial variables are removed from problem
void remove(double** a,int* rhv,int N,int M)
{
   int i;
   for(i=0; i < N - M; i++)
   {
   while(rhv[i] > N)
   {
   int j;
   for(j=i; j < N-1; j++)
```

```
    {
    rhv[j] = rhv[j+1];
    int k;
    for(k=0; k < M+1; k++)
    a[k][j+1] = a[k][j+2];
    }
    }
    }
}

// Final results are displayed
void display(double** a,int* lhv,int M,int N1,int flag,int p)
{
    cout << setprecision(6);          //Format of output is set
    cout << setiosflags(ios::left);
    double *x = new double[N1];
    switch(flag)
    {
    case 0:
      int i;
      for(i=0; i < N1; i++)
      x[i] = 0.0;
      int j;
      for(j=1; j < M + 1; j++)
      {
      int lv = lhv[j - 1];
      if(lv < N1 + 1)
      x[lv - 1] = a[j][0];
      }
      cout << endl << endl << "OPTIMAL SOLUTION :" << endl;
      cout << endl
           << "Objective function = " << setw(10) << a[0][0]*p << endl;
      for(i=0; i < N1; i++)
      cout << endl << "x[" << i + 1 << "] = " << setw(10) << x[i];
      break;
    case 1:
      cout << endl << endl << "CONSTRAINTS INCOMPATIBLE";
      break;
      case 2:
      cout << endl << endl << "OBJECTIVE FUNCTION NOT BOUNDED";
      break;
    }
    delete [] x;
 }
```

Chapter 11

Applications in Administration

11.1 Introduction

In this chapter we give a collection of programs useful in adminstration. In section 11.2 we consider students names and addresses. Inheritance is used. Inheritance is an important concept in object-oriented programming. It enables the spin-off of a new class from an existing one. The new class is the subclass and its parent class is the superclass. In section 11.3 a line editor is implemented applying linked lists. Section 11.4 shows how mergesort can be applied to a list of strings. A linked list of strings is implemented in section 11.5. We can add and delete an element from the list. Furthermore we can display the list. The program is menu driven. A simple case of data encryption is studied in section 11.6. We show how a dictonary can be implenented using a vector class in section 11.7. A simple spell checker is described in section 11.8. In section 11.9 we implement the Huffman code. A simplified version of the *Eliza* program is given in section 11.10. Section 11.11 gives an implementation of histograms (frequency distributions) for letters. For the famous farmer, wolf, goat and cabbage problem two implementations are given in section 11.12. Roman numbers and arabic numbers are treated in section 11.13. The programs can convert roman numbers to arabic numbers and vice versa. The day of the week is calculated in section 11.14 using the Zeller congruence. A class for the Greogorian calender is implemented in section 11.15. The program `records.cpp` in section 11.16 shows how to handle disk files. We use the `fstream` class. A coin machine is implemented as a finite state machine in section 11.17. A program which could be useful in the real estate business is given in section 11.18.

11.2 Names, Address and Inheritance

The following program uses inheritance. This a simple concept that allows the designer to take existing objects and create new one that inherit the properties of the earlier object. The base class in our program is `Name`. This class contains the name `(first,middle,last)` of an individual. The derived class is `Address`. Class `Address` contains the address of the individual. The member function `show_info` in `class Address` uses the member function `show` which belongs to `class Name`.

```
// stud1.cpp

#include <iostream.h>  // for cin(), cout()
#include <string.h>    // for strcpy(), strcat()

   // Declaration of class Name
class Name  {
  private:
    char first[80];
    char mid[80];
    char last[80];
  public:
     Name();
     void set_Name(char*,char*,char*);
     char* show();
};

      // Definition of class Name
Name::Name()
{
   first[0] = mid[0] = last[0] = '\0';
}

void Name::set_Name(char* f, char* m, char* l)
{
   strcpy(first,f);
   strcpy(mid,m);
   strcpy(last,l);
}

char* Name::show()
{
   char* temp = new char[80];
   strcpy(temp,first);
```

```
   strcat(temp," ");
   strcat(temp,mid);
   strcat(temp," ");
   strcat(temp,last);
   strcat(temp,": ");
   return temp;
}

   // Declaration of class Address derived from class Name
class Address : public Name
{
  private:
    char street[80];
    char city[80];
    char state[80];
    char zip[10];

  public:
    Address();
    void new_Address(char*,char*,char*,char*);
    void new_Name(char*,char*,char*);
    char* show_info();
};

   // Definition of class Address
Address::Address()
{
   street[0] = city[0] = state[0] = zip[0] = '\0';
}

void Address::new_Address(char* s1,char* s2,char* s3,char* s4)
{
  strcpy(street,s1);
  strcpy(city,s2);
  strcpy(state,s3);
  strcpy(zip,s4);
}

void Address::new_Name(char* f,char* m,char* l)
{
  set_Name(f,m,l);
}

char* Address::show_info()
{
```

```
  char* temp = show();
  strcat(temp,street);
  strcat(temp," ");
  strcat(temp,city);
  strcat(temp,", ");
  strcat(temp,state);
  strcat(temp,"  ");
  strcat(temp,zip);
  return temp;
}

int main()
{
   Address Black_J_H, Cooper_J_F, Smith_L_J;

   Black_J_H.new_Address("Thomson Street", "Egoli", "TV", "2786");
   Black_J_H.new_Name("Joe", "Henry", "Black");

   Cooper_J_F.new_Address("Main Street", "Pretoria", "NTV", "2490");
   Cooper_J_F.new_Name("Jane", "Floppy", "Cooper");

   Smith_L_J.new_Address("Kings Way", "Witbank", "ETV", "2340");
   Smith_L_J.new_Name("Lucky", "Jack", "Smith");

   cout << Cooper_J_F.show_info() << "\n";

   return 0;
}
```

11.3 Line Editor and Nested Classes

Consider the string of words

```
"Romeo and Juliet by William Shakespeare"
```

Each string of words is a list of words. We represent a sentence as a linked list of words. A word is defined as a character string delimited by a white-space character, a new-line character, or the beginning of a line.

Given the string

```
char *string = "And twenty caged nightingales do sing:"
"or wilt thou sleep ? we'll have thee to a couch";
```

We want to find the n-th word in the sentence. For example the function call

```
x.nth_Token(4)
```

should give:

```
nightingales
```

The function `strtok()` is used extensively. This function searches one string of tokens, which are separated by delimiters in a second string. The program uses the technique of linked lists.

The program can also be used for a text file.

```
// nested.cpp

#include <iostream.h>
#include <stdlib.h>    // for exit()
#include <string.h>    // for strlen(), strcpy(), strtok()

      // Declaration of class Node
class Node
{
   public:
     char* value;
     Node* next;
};

     // Declaration of class Sentence
class Sentence
{
   protected:
      Node* toke;
      int size;
   public:
      Sentence(char*);              // constructor
      ~Sentence();                  // destructor
      char* nth_element(int);       // member function
      void show_all();              // member function
};

  // Definition of class Sentence
Sentence::Sentence(char* o_Sentence)
{
   const char* SPACE = " ";
   char* temp;
   temp = strtok(o_Sentence, SPACE);
   if(temp == NULL)
   {
   cout << "error condition\n";
   size = 0;
   exit(1);
   }
   toke = new Node;
   Node* cursor;
   cursor = toke;
   cursor -> value = new char[strlen(temp) + 1];
   strcpy(cursor -> value, temp);
   cursor -> next = NULL;
```

```
   int i;
   for(i=1; ((temp = strtok(NULL,SPACE)) != NULL); i++)
   {
   cursor -> next = new Node;
   cursor = cursor -> next;
   cursor -> value = new char[strlen(temp) + 1];
   strcpy(cursor -> value, temp);
   cursor -> next = NULL;
   }
   size = i;
}

Sentence::~Sentence()
{
   Node *prev;
   Node *cursor;
   cursor = toke;
   while(cursor != NULL)
   {
   prev = cursor;
   cursor = cursor -> next;
   delete prev -> value;
   delete prev;
   }
}

char* Sentence::nth_element(int num)
{
   Node* cursor;
   cursor = toke;

   int i;
   for(i=1; i < num; i++)
   if(cursor -> next != NULL)
   cursor = cursor -> next;
   else
   return 0;
   return cursor -> value;
}

void Sentence::show_all()
{
   Node* cursor = toke;
   int i;
   for(i=0; i < size; i++) {
```

```
   cout << "==>" << cursor -> value << "\n";
   cursor = cursor -> next;
   }
}     // end definition of class Sentence

int main()
{
   Sentence x("Romeo and Juliet by William Shakespeare");

   x.show_all();
   cout << endl;

   char* c = x.nth_element(3);
   cout << c;                        // output Juliet
   cout << endl << endl;

   Sentence y("If a hart do lack a hind "
              "Let him seek out Rosalind "
              "If the cat will after kind "
              "So be sure will Rosalind ");

   y.show_all();
   cout << endl;
   char* d = y.nth_element(12);
   cout << d;

   return 0;
}
```

The next program is a modification of the program `nested.cpp`.

```
// contain.cpp

#include <iostream.h>
#include <stdlib.h>      // for exit()
#include <string.h>      // for strlen(), strcpy(), strtok()

    // Declaration of class Node
class Node {
  private:
    char* value;
    Node* next;
  public:
    Node(char*);               // constructor one argument
   ~Node();
    void put_next(Node*);
    Node* get_next();
    char* show_value();
};

    // Definition of class Node
Node::Node(char* data)
{
   value = new char[strlen(data) + 1];
   strcpy(value, data);
   next = NULL;
}

Node::~Node() { delete value; }

void Node::put_next(Node* tnode) { next = tnode; }

Node* Node::get_next() { return next; }

char* Node::show_value() { return value; }

    // Declaration of class Token
class Token {
  private:
    Node *toke;
    int size;
  public:
    Token(char*);
   ~Token();
```

```
    char* nth_Token(int);
    void show_all();
};

      // Definition of class Token
Token::Token(char* x)
{
   const char* SPACE = " ";
   char* temp;
   temp = strtok(x,SPACE);
   if(temp == 0) {
   cout << "error condition\n";
   size = 0;
   exit(1);
   }
   toke = new Node(temp);
   Node* cursor;
   cursor = toke;
   temp = strtok(NULL,SPACE);
   int i;
   for(i=1;((temp = strtok(NULL,SPACE)) != NULL); i++)
   {
   cursor->put_next(new Node(temp));
   cursor = cursor->get_next();
   }
   size = i;
}

Token::~Token()
{
   Node *prev;
   Node *cursor;
   cursor = toke;
   while(cursor != NULL)
   {
   prev = cursor;
   cursor = cursor->get_next();
   delete prev;
   }
}

char* Token::nth_Token(int num)
{
   Node* cursor;
   cursor = toke;
```

```
    int i;
    for(i=1 ; i < num ; i++)
    if(cursor != NULL)
    cursor = cursor->get_next();
    else
    return NULL;
    return cursor -> show_value();
}

void Token::show_all()
{
    Node* cursor;
    cursor = toke;
    int i;
    for(i=0;i < size; i++)
    {
    cout << "==>" << cursor->show_value() << "\n";
    cursor = cursor->get_next();
    }
}

int main()
{
  char *string = "And twenty caged nightingales do sing: "
 "or wilt thou sleep ? we'll have thee to a couch ";
  Token x(string);
  x.show_all();
  cout << "Display the 4-rd Token: " << x.nth_Token(4);
  cout << endl << endl;
  cout << "Display all Token: "; cout << endl;
  int j;
  for(j=1;j<18; j++)
  cout << x.nth_Token(j) << "\n";
  return 0;
}
```

11.4 Mergesort Applied to a List

The following program performs a nonrecursive natural mergesort for singly link lists. The problem with quicksort is that it does not always decompose the list into two equal sublists. *Mergesort* just slices the list in two equal halves and sorts the halves separately.

The program uses a linked list to tackle the problem.

Given

```
char* test[] = {
   "so", "smile", "the", "heavens",
   "upon", "this", "holy", "act",
   "that", "after", "hours", "with",
   "sorrow", "chide", "us", "not", NULL
};
```

we find as output

```
[07] act
[09] after
[13] chide
.
.
[01] smile
[00] so
[12] sorrow
.
.
[11] with
```

Thus the list of words is sorted and the words are numbered after their appearance.

```
// mergelist.cpp

#include <stdio.h>
#include <string.h>
#include <stdlib.h>

class Node {
   public:
      char* s;
      int sequence;
      Node *next;
};

void WriteList(Node *l)
{
   Node *node;
   for(node = l; node; node = node->next) {
      printf("[%02d] %s\n", node->sequence, node->s);
   }
}

typedef int (*cmpFunc_t)(Node *a,Node *b);

int Compare(Node *a,Node *b)
{
   return strcmp(a->s, b->s);
}

Node *Merge(Node *a,Node *b,cmpFunc_t cmpFunc)
{
   Node *c, *ctail;
   if(!a)
   {  return b; }
   else if(!b)
   {  return a; }
   if(cmpFunc(a, b) <= 0) {
   c = ctail = a;
   a = a -> next;
   } else {
   c = ctail = b;
   b = b->next;
   }
   while(a && b) {
   if(cmpFunc(a,b) <= 0) {
   ctail = ctail->next = a;
```

```
a = a->next;
} else {
ctail = ctail->next = b;
b = b->next;
}
}
if(!a) {
ctail->next = b;
} else {
ctail->next = a;
}
return c;
}

Node *NaturalListsort(Node *l,cmpFunc_t cmpFunc)
{
Node *runs[32];
Node *x, *r, *rtail, *prev, *next;
int depth, maxDepth;

if(!l || !l->next)
{ return l; }
for(depth=0;depth<sizeof(runs)/sizeof(runs[0]);depth++) {
runs[depth] = 0;
}
maxDepth = 0;  prev = l;
x = prev->next;
r = rtail = prev;
rtail -> next = 0;
while(x) {
next = x -> next;
if (cmpFunc(prev,x) > 0) {
depth = 0;
while(runs[depth]) {
r = Merge(runs[depth],r,cmpFunc);
runs[depth] = 0;
depth++;
}
if(depth > maxDepth) {
maxDepth = depth;
}
runs[depth] = r;
r = rtail = x;
} else {
rtail = rtail->next = x;
```

```
    }
    prev = x;
    x = next;
    rtail->next = 0;
    }
    for(depth = 0; depth <= maxDepth; depth++) {
    if(runs[depth]) {
    r = Merge(runs[depth],r,cmpFunc);
    }
    }
    return r;
}

Node *NewNode(char *s)
{
    static int seq = 0;
    Node *node;
    node = new Node;
    node->s = new char[strlen(s)+1];
    strcpy(node->s,s);
    node->sequence = seq++;
    node->next = 0;
    return node;
}

char *test[] = {
    "so", "smile", "the", "heavens", "upon", "this",
    "holy", "act", "that", "after", "hours", "with",
    "sorrow", "chide", "us", "not", NULL
};

void main()
{
    Node *l, *ltail;
    char **s;
    l = ltail = NewNode(*test);
    for(s = test + 1; *s; s++) {
    ltail = ltail->next = NewNode(*s);
    }
    l = NaturalListsort(l,Compare);
    WriteList(l);
}
```

11.5 Linked List of Strings

The following menu driven program processes a linked list of strings. We can create a list of strings. Then we can add and delete elements of the list.

```
// linked.cpp

#include <iostream.h>
#include <stdio.h>
#include <stdlib.h>
#include <string.h>

class List_element
{
   public:
    char item[40];             // data item for this node
    List_element* next;        // pointer to the next node
};

typedef class List_element node;     // class type declaration

int menu();
void create(node*);
node* insert(node*);
node* locate(node*,char[]);
node* scrap(node*);
void display(node*);

void main()
{
     node* start;               // pointer to beginning of list
     int choice;

     do
     {
     choice = menu();
     switch(choice)
     {
     case 1:
        start = new node;  // allocate space for first node
        create(start);
        cout << "\n";
        display(start);
```

```
        continue;

    case 2:
        start = insert(start);
        cout << "\n";
        display(start);
        continue;

    case 3:
        start = scrap(start);
        cout << "\n";
        display(start);
        continue;

    default:
        cout << "End of computation\n";
    }
    }
    while(choice != 4);
}

int menu()
{
    int choice;
    do
    {
    printf("\n MENU: \n");
    printf("  1 - CREATE the linked list\n");
    printf("  2 - ADD a component\n");
    printf("  3 - DELETE a component\n");
    printf("  4 - END\n");
    printf("Please enter your choice (1, 2, 3 or 4) -> ");
    cin >> choice;
    if(choice < 1 || choice > 4)
    printf("\nERROR - Please try again\n");
    }
    while(choice < 1 || choice > 4);
    cout << "\n";
    return choice;
}

void create(node *record)        // create a linked list
{
    printf("Data item (type \'END\' when finished): ");
    scanf(" %[^\n]", record->item);
```

```
    if(strcmpi(record->item, "END") == 0)
            record->next = NULL;
    else
    {
    record->next = new node;
    create(record->next);
    }
}

void display(node *record)      // display the linked list
{
    if(record->next != NULL)
    {
    printf("%s\n",record->item);
    display(record -> next);
    }
}

node *insert(node *first)  // add one component to the linked list
{
    node *newrecord;            // pointer to new node
    node *tag;                  // pointer to node before target node
    char newitem[40];           // new data item
    char target[40];            // data item following the new entry
    cout << "New data item: ";
    scanf(" %[^\n]", newitem);
    printf("Place before (type /'END/' if last): ");
    scanf(" %[^\n]", target);
    if(strcmpi(first->item, target) == 0)
    {
    newrecord = new node;
    strcpy(newrecord->item, newitem);
    newrecord->next = first;
    first = newrecord;
    }
    else  {
    tag = locate(first, target);
    if(tag == NULL)
    printf("\nMatch not found - Please try again\n");
    else  {
    newrecord = new node;
    strcpy(newrecord->item, newitem);
    newrecord->next = tag->next;
    tag->next = newrecord;
```

```
    }
    }
    return first;
}

node *locate(node *record,char target[])
{
    if(strcmpi(record->next->item,target) == 0)
    return record;
    else
    if(record -> next -> next == NULL)
    return NULL;
    else
    return locate(record->next,target);
}

node *scrap(node *first)
{
    node *tag;
    node *temp;
    char target[40];
    printf("Data item to be deleted: ");
    scanf(" %[^\n]", target);
    if(strcmpi(first->item, target) == 0)
    {
    temp = first->next;
    delete first;
    first = temp;
    }
    else
    tag = locate(first, target);
    if(tag == NULL)
    printf("\nMatch not found - Please try again\n");
    else {
    temp = tag->next->next;
    delete tag->next;
    tag->next = temp;
    }
    return first;
}
```

11.6 Data Encryption

The basic string operations of comparing, finding a substring, finding the length, finding the position of a character, concatenating, copying, inserting, deleting etc. are adequate for most applications that use strings. In *data encryption* the basic operations are replacement, permutation and swapping. *Encryption* refers to the coding of information in order to keep it secret. The string of characters describing the information is transformed (mapped) into another string. A simple string class is used.

```
// cipher.cpp

#include <iostream.h>
#include <string.h>
#include <ctype.h>   // for isprint

    // Declaration of class String
class String
{
   private:
      char *s;                  // pointer to string contents
      int length;               // length of string in characters
   public:
      String(char *text);       // constructor using existing string
      String(int);              // creates default empty string
      String(const String&);    // copy constructor
      ~String();                // destructor
      String operator + (const String&);
      String& operator = (const String&);   // assignment operator
      void concat(const String&,const String&);
      void display();
      String caesar(String,int);
};

    // Definition of class String
String::String(char *text)
{
   length = strlen(text);
   s = new char[length+1];
   strcpy(s,text);
}

String::String(int size)
{
```

```
   length = size;     // get length of string
   s = new char[length+1];
   s[length] = '\0';
}

String::String(const String& other)
{
   length = other.length;         // length of other string
   s = new char[length+1];        // allocate the memory
   cout << "copy constructor invoked " << endl;
   strcpy(s,other.s);             // copy the text
}

String::~String()
{
   delete s;
}

String String::operator + (const String& arg)
{
   String temp(length + arg.length);
   strcpy(temp.s, s);
   strcat(temp.s, arg.s);
   return temp;
}

String& String::operator = (const String &arg)
{
   if(this == &arg)
   { return *this; }
   delete [] s;
   s = new char[strlen(arg.s) + 1];
   strcpy(s, arg.s);
   cout << "assignment invoked " << endl;
   return *this;
}

void String::concat(const String& s1,const String& s2)
{
   delete s;                                // forget current string
   length = strlen(s1.s) + strlen(s2.s);
   s = new char[length];                    // allocate the memory
   strcpy(s,s1.s);                          // copy first into space
   strcat(s,s2.s);                          // add the second
}
```

```
void String::display()
{
   cout << s;
}

String String::caesar(String A,int k)
{
   const int first = 32;      // printable range is
   const int printable = 95;  // 32 through 126

   delete s;
   length = strlen(A.s);
   s = new char[length+1];
   strcpy(s,A.s);

   while(k < 0)
   k += printable;

   int i;
   for(i=0; i < length; i++)
   if(isprint(s[i]))
   s[i] = ((((s[i]-first)+ k)%printable) + first);

   return s;
}

int main()
{
   cout << endl;
   char* test = "the ideas of march";
   String S(test);
   S.display();
   cout << endl;
   S.caesar(test,3).display();

   return 0;
}
```

11.7 Vector Class and a Dictionary

Here we apply the vector class to implement a dictionary. The input is a name. If the name is in the vector the output is the years when the person lived. If the name is not in the vector we obtain the message:

```
name not in list
```

```
// dicto.cpp
// Application of class templates

#include <iostream.h>
#include <string.h>    // for strcpy, strcmp
#include <stdio.h>     // for gets

      // declaration of class Vector
template <class T> class Vector
{
  protected:
    T *data;
    int size;

  public:
    Vector(int);             // constructor one argument
    ~Vector();               // destructor
    T& operator[](int);      // overloading the operator []
    friend void database();
};

// Implementation of class Vector
template <class T> Vector<T>::Vector(int n)
{
   data = new T[n];
   size = n;
}

template <class T> Vector<T>::~Vector()
{
   delete[] data;
}

template <class T> T& Vector<T>::operator[](int i)
{
   return data[i];
}
```

```
int main()
{
  const int N = 5;

  Vector<char*> years(N);
  Vector<char*> names(N);
  years[0] = "1777-1855";   names[0] = "Gauss";
  years[1] = "1646-1716";   names[1] = "Leibniz";
  years[2] = "1643-1727";   names[2] = "Newton";
  years[3] = "1826-1866";   names[3] = "Riemann";
  years[4] = "1802-1829";   names[4] = "Abel";
  cout << endl << endl;
  char name[10];

  cout << "enter name: ";
  gets(name);

  int test[N];
  int v;
  int i;
  for(i=0; i<N; i++)
  {
  v = i;
  test[i] = strcmp(name,names[i]);
  if(test[i] == 0) break;
  else
  if((i==(N-1)) && test[i] != 0)
  cout << name << " : name not in list";
  }
  if(test[v] == 0)
  cout << name << ": " << years[v];

  return 0;
}
```

11.8 Spelling Checker

We develop a simple spelling checker program to illustrate the use of the linked-list, linkedlist-iterator and string classes. The names of the input and output files are supplied via a command-line parameters. Hence

```
spellchk inn.ext outn.ext
```

runs the spelling checker on the file `inn.ext` and writes the output onto `outn.ext`.

The dictionary is read from file into a linked list of strings. The text file is read line for line and each line is split up into words. The function `InDictionary` is called for each word. It performs a linear search to check whether the word is in the dictionary. This would of course be unacceptably slow for commercial spelling checkers. If the word is found the next word is checked. On the other hand, if the word is not found in the dictionary the program allows you to either add the word to the dictionary, to edit the word, or to ignore the word leaving the text and the dictionary unaltered.

```
// spellchk.cpp

#include <fstream.h>
#include <iostream.h>
#include <process.h>
#include <conio.h>

#include "string.cpp"
#include "vector.cpp"
#include "linklist.cpp"

LinkedList<string> dictionary;

int changed = 0;

void LoadDictionary()
{
  cout << "Loading Dictionary ..." << endl << endl;
  string word;
  ifstream dictfile ("Dict.Dat");
  if(dictfile)
  {
  long int count = 0;
  while(!dictfile.eof())
  {
  dictfile >> word;
```

```
 dictionary.push(word);
 ++count;
 }
 dictionary.pop();
 --count;
 cout << "Read " << count << " words from dictionary." << endl;
 }
 else
 {
 cout << "*** ERROR ***: cannot find dictionary file" << endl;
 exit(0);
 }
 dictfile.close();
};

void SaveDictionary()
{
  cout << "Saving Dictionary ..." << endl << endl;
  ofstream dictfile("Dict.Dat");

  if(!dictfile)
  {
  cout << "*** ERROR ***: cannot open dictionary file" << endl;
  exit(0);
  };

  LinkedListIterator<string> dictword(dictionary);

  long int count = 0;
  for(dictword.reset(); ~dictword; ++dictword)
  {
  dictfile << dictword() << endl;
  ++count;
  }
  cout << "Saved " << count << " words to dictionary file." << endl;
  dictfile.close();
};

int InDictionary(const string& word)
{
  LinkedListIterator<string> dictword(dictionary);
  dictword.reset();
  while((~dictword) && (!word.CaselessCompare(dictword())))
    ++dictword;
  return (~dictword);
```

```
};

void SpellCheck(ifstream& infile,ofstrean& outfile)
{
  string line;
  while(!infile.eof())
  {
  infile >> line;

  int first = 0, last, length = line.Length();
  while(first < length)
  {
  while((first < length) && (!isalpha(line[first])))
  first++;
  if(first < length)
  {
  last = first + 1;
  while((last < length) && (isalpha(line[last])))
  ++last;

  string word = line.substring(first,last - first);

  if(!InDictionary(word))
  {
  cout << "The word  [" << word
       << "] was not found in dictionary." << endl << endl;
  cout << "Do you want to " << endl
       << "       A      ->  Add to dictionary?" << endl
       << "       E      ->  Edit the word?"     << endl
       << "   <ENTER>    ->  Ignore the word?"   << endl;
  char choice = toupper(getch());
  switch(choice)
  {
  case 'A': dictionary.append(word);
  changed = 1;
  break;
  case 'E': cout << endl;
  int oldlength = word.Length();
  cout << "Enter correct spelling for " << word
       << ": " << endl;
  cin >> word;
  int diflength = word.Length() - oldlength;
  line = line.substring(0,first) + word +
         line.substring(last,length-last);
  last   += diflength;
```

```
  length += diflength;
  break;
  };
  };
  first = last + 1;
  };
  };
  outfile << line << endl;
  };
};

void main(int argc,char* argv[])
{
  if(argc < 3)
  {
  cout << "*** ERROR ***,  usage : spellchk <InFileName> <OutFileName>"
       << endl;
  exit(0);
  }

  ifstream infile(argv[1]);
  if(!infile)
  {
  cout << "** ERROR **: cannot find input file " << argv[1] << endl;
  exit(0);
  }
  ofstream outfile(argv[2]);
  if(!outfile)
  {
  cout << "** ERROR **: cannot open output file " << argv[1] << endl;
  exit(0);
  }
  LoadDictionary();
  SpellCheck (infile,outfile);
  if(changed)
  {
  cout << "Do you want to save the changes to the dictionary? (Y/N)";
  char answer = toupper(getch());
  if(answer == 'Y')  SaveDictionary();
  };
  infile.close();
  outfile.close();
};
```

11.9 Huffman Code

Data compression can be achieved by assigning short codes to the most frequently encountered source characters and necessarily longer codes to the others. The Morse code is such a variable length code which is reasonably efficient for both English and German text. The most frequently encountered letter, "E", has the shortest Morse code ("·") while the less frequently used letters (e.g. "Q") have longer codes (e.g. "− − · −").

The Morse code is a punctuated code in that after each letter there is a letter space of 3 time units and after each word there is a word space of 6 time units. This introduces a significant deterioration of efficiency.

One requirement for a code to be generally useful is that it should be uniquely decodable. A code is said to be uniquely decodable if, for each code-word sequence of finite length, the resulting string of letters of the code alphabet is different from any other codeword sequence.

Consider for example a code defined on a binary alphabeth $\mathcal{C} = \{0, 1\}$. A code C with codewords

$$c \in \{0, 1, 00, 01, 10, 11\}$$

is for example not uniquely decodable since the string "01" could be decoded as either the concatenation of 2 codewords [0]+[1] or as the single code word [01]. On the other hand, a code with codewords

$$c \in \{0, 10, 11\}$$

is uniquely decodable. Try and form a string with these codewords which is not uniquely decodable.

A class of very useful codes are the prefix or instantaneous codes. A code is called a prefix code if no code-word is a prefix of any other code-word. Prefix codes are uniquely decodable. Hence, from the Kraft inequality we know that the expected code length is greater or equal to the entropy of the source. Consider for example a source (e.g. the text file to be compressed) of four possible output symbols

$$x \in \mathcal{X} = \{a, b, c, d\}$$

Assume the probabilities for the four symbols are

$$\{p(x)\} = \left\{\frac{1}{2}, \frac{1}{4}, \frac{1}{8}, \frac{1}{8}\right\}.$$

We can construct a prefix code C with $M = 4$ codewords

$$c \in \mathcal{C} = \{0, 10, 110, 111\}.$$

Note that this code is uniquely decodable and furthermore, there is no need to add punctuating symbols between the code-words. Prefix codes are self-punctuating. The average length for this code is

$$L(C) = \sum_{m=1}^{4} p_m l_m = \frac{1}{2} \cdot 1 + \frac{1}{4} \cdot 2 + 2 \cdot \frac{1}{8} \cdot 3 = 1\frac{3}{4}\,.$$

A prefix code based on an alphabet of size $s_{\mathcal{A}}$ can be represented graphically by an s-ary tree for a binary tree) where each node has $s_{\mathcal{A}}$ branches.

Each node of the tree for this code has two branches since the code is based on a binary alphabet. The code-words are represented by the leaves of the tree. Since no code-word is a prefix of any other code-word, each code-word eliminates its descendants (nodes which are not a leaf).

Huffman codes are optimal in the sense that its expected length is equal or shorter than the expected length of any other uniquely prefix code. Huffman codes are constructed by combining at each step the s least probable source symbols into a new source symbol with the combined probability and inserting into each code-word string one of the s letters of the alphabet.

For example, consider a source of four symbols

$$\mathcal{X} = \{\, A, B, C, D \,\}$$

with probability distribution

$$\{p_m\} = \left\{ \frac{1}{2}, \frac{1}{4}, \frac{1}{8}, \frac{1}{8} \right\}.$$

We construct a binary Huffman code for this source. We first combine the two least probable symbols C and D pushing the binary letters 0 and 1 into the code-word strings. The joint probability is $\frac{1}{4}$. This process is repeated by combining repetitively the two least probable symbols.

The following program reads in the number of different source symbols (e.g. the ASCII symbols in the case of a text file) and produces a binary Huffman code for this source.

```
// Huffman.cpp

#include <process.h>
#include <iostream.h>


const int maxsymbols = 50; // maximum number of symbols
  // in source alphabet.

float pm[maxsymbols+1];  // array for probability distribution
                         // of source.
int M;                   // no of source symbols.

void main()
{
  cout << "M = ";   // read in number of source symbols and
  cin  >> M;        // make sure that M > 2 and M < maxsymbols

  if((M < 2) || (M > maxsymbols))
  {
  cerr << "*** ERROR ***: invalid no of symbols for source"
       << endl;
  abort();
  }

  float sumpm = 0.0;     // used to check that sum_m p_m = 1

  int m;
  for(m=1; m <= M; ++m)
  {
  cout << "p_" << m << " = ";          // and check that
  cin  >> pm[m];                       // 0 <= pm <= 1
  if((pm[m] < 0.0) || (pm[m] > 1.0))
  {
  cerr << "*** ERROR ***: invalid p_m" << endl;
  abort();
  }
  sumpm += pm[m];
  }

  float eps = 0.001;  // check that sum_m p_m = 1
  if((sumpm < 1.0-eps) || (sumpm > 1+eps))
  {
  cerr << "*** ERROR ***: probabilities dont add up to 1.0"
       << endl;
```

```
  abort();
}

struct node
{
  float prob;
  char  digit;
  node* next;
};

// tree always stores the lowest level of node in the tree,
// treep moves level for level up the tree
node *tree[maxsymbols+1], *treep[maxsymbols+1];

for(m=1; m<=M; ++m)
{
   tree[m] = new node;
   tree[m]->prob  = pm[m];
   tree[m]->digit = '-';
   tree[m]->next  = 0;
   treep[m] = tree[m];
}

int   Mp = M;
float min1, min2;
int   mmin1, mmin2;

while(Mp >= 2)
{       // find 2 lowest probabilities
   min1 = min2 = 1.0;    // in current level of tree.
   int m;
   for(m=Mp; m>= 1; --m)
   {
   if(treep[m]->prob < min1) {
   mmin1 = m;
   min1  = treep[m]->prob;
   }
   }
   for(m=Mp; m>= 1; --m)
   {
   if((treep[m]->prob<min2) && (treep[m]->prob>=min1) && (m!=mmin1))
   {
   mmin2 = m;
   min2  = treep[m] -> prob;
   }
```

```
    }

    // now create a new node in the tree for the combined
    // symbol of the two least probable source symbols.
    node* dumnode  = new node;
    dumnode -> prob  = treep[mmin1] -> prob + treep[mmin2] -> prob;
    dumnode -> digit = ' ';
    dumnode->next = 0;

    if(mmin1<mmin2)
    {                    // for consistency in codes make
    int idum = mmin2;    // certain that mmin1 is always
    mmin2 = mmin1;       // the larger index.
    mmin1 = idum;
    }

    treep[mmin2] -> digit = '0';       // set the digit for the 2
    treep[mmin2] -> next  = dumnode;   // symbols which have been
    treep[mmin1] -> digit = '1';       // combined into a single
    treep[mmin1] -> next  = dumnode;   // artificial symbol

    // Now build the next level of tree
    int mpp;
    if(mmin1 < mmin2)
    mpp = mmin1;
    else
    mpp = mmin2;

    treep[mpp] = dumnode;

    int mp=1;
    for(m=1; m<=Mp; ++m)
    {
    if((m != mmin1) && (m != mmin2))
    {
    if(mp == mpp) ++mp;
    treep[mp] = treep[m];
    ++mp;
    }
    }
    --Mp;
}


// Now the full tree is constructed and we only have to
```

```
// read of the code-word strings in reverse order.

char codeword[maxsymbols+1];

node* nd;

cout << endl;
cout << "  m  |          p_m        |   code-word" << endl;
cout << "========================================="
     << endl;

for(m=1; m<=M; ++m)
{
  nd = tree[m];
  cout.width(3);
  cout << m << "  |  ";
  cout.width(15);
  cout << pm[m] << "  |  ";
  codeword[1] = nd->digit;
  int mp = 1;

  while(nd->next != 0)
  {
  ++ mp;
  nd = nd -> next;
  codeword[mp] = nd -> digit;
  }
  int mpp;
  for(mpp=mp; mpp>=1; --mpp)
  cout << codeword[mpp];
  cout << endl;
  };
  char dummy[80];
  cin >> dummy;
};
```

11.10 Eliza

Joseph Weizenbaum showed that a digital computer can carry on an "intelligent" discussion quite well if the realm of discourse is highly restricted. Weizenbaum wrote a program named *Eliza*, which operated as a psychiatrist listening to a patient. Although his program is more complex and involved (and thus a better psychiatrist), the essence of what he did can be captured in a C (C++) program of rather modest size. To return to the program enter bye.

A dialogue would be like this:

```
computer: how are you ?
patient: fine
computer: Tell me more ...
patient: why
computer: remember, therapy is good for you
patient: I am unhappy
compter: why are you unhappy ?
patient: bye
computer: Your bill will be mailed to you
```

The program listing is

```
// eliza.cpp

#include <iostream.h> // for cout
#include <stdio.h>    // for gets
#include <string.h>   // for strcmp, strlen
#include <ctype.h>    // for tolower

const int Max = 100;
const int Short = 10;
const int Veryshort = 3;

void respond(char* s);
int lookup(char* token);
void assert_topic(char* t);
int find_topic(char* t);
int in_topics(char* s);
void get_token();
int is_in(char c,char* s);

char *response[] = {
  "how are you ? ",
  "Is it not a wonderful day ? ",
```

```
  "Did you have a happy childhood ?",
  "Did you hate your father ? ",
  "Did you hate your mother ? ",
  "I'm not sure I understand. ",
  "Do you pray ? ",
  ""
};

char *trans[] = {
  "you", "Let's not talk about me.",
  "love", "We must love God.",
  "family", "Do you come from a close-knit family ?",
  "think", "Why do you think that ?",
  "hate", "So you hate something - tell me more.",
  "what", "Why do you ask ?",
  "want", "Why do you want that ?",
  "need", "We all need many things - is this special ?",
  "why", "Remember, therapy is good for you.",
  "know", "How do you know that ?",
  "bye", "Your bill will be mailed to you.",
  "murder", "I don't like killing.",
  "kill", "It is wrong to kill.",
  "jerk", "Don't ever call me a jerk!",
  "can't", "Don't be negative - be positive.",
  "relax", "Try to relax.",
  "failure", "Strive for success.",
  "never", "Don't be negative - positive",
  "unhappy", "Why are you unhappy ?",
  "happy", "Why are you happy ?",
  "liqueur", "Liqueur is not good for your health",
  ""
};

char topics[Max][80];
char token[80];
char* p_pos;
int res = 0;
int head = 0;
int tail = 0;

int main()
{
  char s[80];
  cout << response[res++];
  do
```

```
  {
  cout << " ";
  p_pos = s;
  gets(s);
  respond(s);
  } while(strcmp(s,"bye"));
  return 0;
}

void respond(char* s)
{
  char t[80];
  int loc;
  if(strlen(s) < Veryshort && strcmp(s,"bye"))
  {
  if(find_topic(t))
  {
  cout << "You just said: ";
  cout << t;  cout << endl;
  cout << "tell me more" << endl;
  }
  else
  {
  if(!*response[res]) res = 0;
  cout << response[res++];
  }
  return;
  }
  if(in_topics(s))
  {
  cout << "Stop repeating yourself!" << endl;
  return;
  }
  if(strlen(s) > Short) assert_topic(s);
  do
  {
  get_token();
  loc = lookup(token);

  if(loc != -1)
  {
  cout << trans[loc+1] << endl;
  return;
  }
  } while(*token);
```

```
  cout << "Tell me more..." << endl;
}

int lookup(char* token)
{
  int t;
  t = 0;
  while(*trans[t])
  {
  if(!strcmp(trans[t],token)) return t;
  t++;
  }
  return -1;
}

void assert_topic(char *t)
{
   if(head == Max) head = 0;
   strcpy(topics[head],t);
   head++;
}

int find_topic(char* t)
{
  if(tail != head)
  {
  strcpy(t,topics[tail]);
  tail++;
  if(tail == Max) tail=0;
  return 1;
  }
  return 0;
}

int in_topics(char* s)
{
  int t;
  for(t=0; t<Max;t++)
  if(!strcmp(s,topics[t])) return 1;
  return 0;
}

void get_token()
{
   char* p;
```

```
   p = token;
   while(*p_pos==' ') p_pos++;
   if(*p_pos == '\0')
   {
   *p++='\0';
   return;
   }
   if(is_in(*p_pos,".!?"))
   {
   *p = *p_pos;
   p++, p_pos++;
   *p = '0';
   return;
   }
   while(*p_pos!=' ' && !is_in(*p_pos,".,;?!") && *p_pos)
   {
   *p = tolower(*p_pos++);
   p++;
   }
   *p = '\0';
}

int is_in(char c,char* s)
{
  while(*s)
  {
  if(c==*s) return 1;
  s++;
  }
  return 0;
}
```

11.11 Counting the Letters in a Phrase

Given the phrase

```
Yes, by all means; if it be but to scrape the figures out of your
husband's brains
```

We would like to find the number of a's b's c's etc. in the given sentence. The sentence above, for example, includes 4 a's and 6 s's. The following program counts the number of times the 26 different letters of the alphabet appear.

The program `alph1.cpp` uses the the function `strchr`. The function `strchr` scans a string in the forward direction, looking for a specific character. `strchr` finds the first occurrence of the character *c* in the string.

In the program `alph2.cpp` we implement the counting of the letters as a function. The function `lower` checks if all characters are lowercase letters and returns 0 if there is a character that is not a letter or is not in lower case. It returns 1 if all characters are lower case letters.

In the program `histoc.cpp` we give a graphical output of the result. We also use the class concept.

```
// alph1.cpp

#include <iostream.h>
#include <stdio.h>
#include <string.h>   // for strchr
#include <conio.h>    // for getche

int main()
{
   int alpha[26];
   int c, i;
   char alphalist[26] = "abcdefghijklmnopqrstuvwxyz";
   char *index;    // index into alphalist

   for(i=0;i<26;++i) alpha[i] = 0;

   cout << "enter string: ";

   while((c = getche()) != '\r')
   if((index = strchr(alphalist, c)) != NULL)
   ++alpha[index-alphalist];

   for(i=0;i<26;i++)
   printf("%c occurred %d times.\n",alphalist[i], alpha[i]);

   return 0;
}
```

```
// alph2.cpp

#include <iostream.h>
#include <string.h>   // for strlen
#include <ctype.h>    // for isalpha

int lower(char *s)
{
   int i;
   for(i=0; i < strlen(s); ++i)
   if(!isalpha(s[i]) || !islower(s[i]))
   return 0;
   return 1;
}
```

```
void count(char *s)
{
   int letter[26];
   int i;
   for(i=0; i < 26; i++)
   letter[i] = 0;
   for(i=0; i < strlen(s); i++)
   letter[int(s[i]) - 97] += 1;
   for(i=0; i < 26; i++)
   if(letter[i] != 0)
   cout << "String has " << letter[i] << " '" << char(i+97) << "'.\n";
}

int main()
{
   int length;
   cout << "this program counts the number of each letter in a ";
   cout << "given string " << endl;
   cout << "enter length of string: ";
   cin >> length;
   char *s = new char[length+1];
   do
   {
   cout << "the characters in the string must be lower case letters";
   cout << endl;
   cout << "enter string: ";
   cin >> s;
   } while(lower(s) == 0);

   cout << "\nGiven string is " << s << endl;
   count(s);
   delete s;

   return 0;
}
```

In the following program we give a graphical output of the counting.

```
// histoc.cpp

#include <iostream.h>
#include <ctype.h>     // for islower, toupper, isupper

const int NO_OF_LETTERS = 26;
const int DEFAULT_HEIGHT = 14;

class Histogram
{
   public:
      Histogram();
      void reset();
      void add_char(const char);
      void scale(const int);
      void display() const;
   private:
      int the_letter_freq[NO_OF_LETTERS];
      int the_max_height;
      int the_actual_max;
};

Histogram::Histogram()
{
   reset();
}

void Histogram::reset()
{
   int ch_pos;
   for(ch_pos=0; ch_pos < NO_OF_LETTERS; ch_pos++)
   {
   the_letter_freq[ch_pos] = 0;
   }
   the_max_height = DEFAULT_HEIGHT;
   the_actual_max = 0;
}

void Histogram::add_char(const char character_to_add)
{
   char c = character_to_add;
   if(islower(c))
   {
```

```
   c = toupper(c);
   }
   if(isupper(c))
   {
   c = (char)((int) c - (int) 'A');
   the_letter_freq[c]++;
   }
}

void Histogram::scale(const int height)
{
   int max = 0;
   int ch;
   the_max_height = height;

   for(ch=0; ch<NO_OF_LETTERS; ch++) {
   if(the_letter_freq[ch] > max) {
   max = the_letter_freq[ch];
   }
   }

   the_actual_max = max;

   if(max > 0) {
   for(ch=0; ch<NO_OF_LETTERS; ch++) {
   int scaled = (the_letter_freq[ch]*height)/max;
   int rem = (the_letter_freq[ch]*height)%max;
   the_letter_freq[ch] = (rem + rem > max ? scaled+1 : scaled);
   }
   }
}

void Histogram::display() const
{
   int j;
   for(j=the_max_height; j>0; j--) {
   cout << " | ";
   int ch;
   for(ch_pos=0; ch_pos<NO_OF_LETTERS; ch_pos++) {
      if(the_letter_freq[ch_pos] >= j)
      cout << "*";
      else
      cout << " ";
   }
   cout << " | " << "\n";
```

```
  }

  cout << "    ---------------------------" << "\n";
  cout << "    ABCDEFGHIJKLMNOPQRSTUVWXYZ       * = ";

  if(the_max_height > 0)
    cout << (float) the_actual_max/(float) the_max_height;
  else
    cout << "0.0";
  cout << "\n";
}

int main()
{
  Histogram hist;
  char ch;

  hist.add_char('A');   hist.add_char('C');
  hist.add_char('Z');   hist.add_char('c');
  hist.add_char('C');   hist.add_char('z');

  hist.scale(12);
  hist.display();
  Histogram my;
  char name[] = { 'W', 'i', 'l', 'l', 'i', '\0' };
  char c;

  int i;
  for(i=0; i<5; i++)
  my.add_char(name[i]);
  my.scale(12);
  my.display();

  Histogram yours;
  char text[] = "uniontrustfonds";
  char d;

  int j;
  for(j=0; j<15; j++)
  yours.add_char(text[j]);
  yours.scale(12);
  yours.display();

  return 0;
}
```

11.12 The Farmer, Wolf, Goat and Cabbage Problem

Consider the classic "farmer, wolf, cabbage, and goat" problem. A farmer finds himself on the south side of the Limpopo River holding a wolf, a cabbage, and a goat. Somehow he must transport all four to the north side of the river. Unfortunately, he has only one very small boat. So small, in fact, that in addition to himself it can hold only one other occupant. Only the farmer can row the boat. He cannot leave the goat alone with the cabbage, nor can he leave the wolf alone with the goat. The wolf, fortunately, is a committed carnivore and will not eat the cabbage. What are the crossings the farmer must perform to carry all four to the north side of the Limpopo ?

To solve this problem, we first need a way of encoding the positions of the various players. A convenient representation is to use numbers between 0 and 15, interpreting the four binary digits in each number as four bit values. We can use a 0 to represent south, and a 1 value to represent north. The four bits in each number will represent the locations of the farmer, the wolf, the cabbage, and the goat, respectively. Thus the number 5 will be interpreted as representing the fact that the farmer and the cabbage are on the south side of the Limpopo, while the wolf and goat are on the north side. A most unsafe position, and one that will probably quickly change to 4 when the wolf eats the goat. Another option is to introduce an array `int v[4]`, where `v[0]` denotes the farmer, `v[1]` denotes the wolf, `v[2]` denotes the cabbage, and `v[3]` denotes the goat. If the farmer is on the south side `v[0]` takes the value 0. If the farmer is on the north side `v[0]` takes the value 1 (likewise for the others).

We set

```
v[0] = farmer, v[1] = wolf, v[2] = cabbage and v[3] = goat
```

This program uses the abbreviations F for Farmer, W for wolf, C for Cabbage and G for Goat. A 0 or 1 is allocated to each depending on which side of the river they are. The program finds a way from position 0000 where everyone is on the one side to 1111 where everyone is on the other side. This program only gives the first solution it finds. In reality there are two solutions to the problem. We give two implementations. The program `river.cpp` uses the class concept.

```
// farmer.cpp

#include <iostream.h>

int safe(int v[])
{
   if((v[3] == v[2]) && (v[3] != v[0]))  // == equality operator
   return 0;

   if((v[3] == v[1]) && (v[3] != v[0]))  // && logical AND operator
   return 0;

   return 1;  // anything else is safe
}

int cross(int s)
{
   s = s + 1;
   if(s == 2) s = 0;
   return s;
}

int complete(int v[])
{
   int c = 0;
   if((v[0] == 1) && (v[1] == 1) && (v[2] == 1) && (v[3] == 1))
   c = 1;
   return c;
}

int main()
{
   int v[4], w[4];
   v[0] = 0;  v[1] = 0;  v[2] = 0;  v[3] = 0;
   cout << endl << "v[0] = " << v[0] << "   v[1] = " << v[1]
        << "   v[2] = " << v[2] << "   v[3] = " << v[3];
   int prev = 6;
   do
   {
   int count = 5;
   int i;
   for(i=1;i<4; i++)
   {
   int j;
   for(j=0; j<4; j++)
```

```
    {
    w[j] = v[j];
    }
    w[0] = cross(v[0]);  w[i] = cross(v[i]);
    if((v[i] == v[0]) && (i != prev) && (safe(w) == 1))
    count = i;
    }
    prev = count;
    w[0] = cross(v[0]);
    int j;
    for(j=1; j<4; j++)
    {
    w[j] = v[j];
    }
    if((v[0] == 1) && (safe(w) == 1)) count = 5;
    if(count == 5)
    {
    v[0] = cross(v[0]);  prev = 5;
    }
    else
    {
    v[0] = cross(v[0]);  v[count] = cross(v[count]);
    prev = count;
    }
    cout << endl << "v[0] = " << v[0] << "   v[1] = " << v[1]
         << "   v[2] = " << v[2] << "   v[3] = " << v[3];
    }
    while (complete(v) == 0);

    return 0;
}
```

```
// river.cpp

#include <iostream.h>

const int n = 4;
const int con1 = 10;
const int con2 = 15;
const int con3 = 20;

int comb[con1][con2];

class Search
{
   private:
     int safe(int v[]);
     int trc[20];
   public:
     void cmbgen();
     void follow();
     void set();
     void sequ();
};

void Search::set()
{
   int i, j;
   for(i=0;i<con1;i++)
   {
   for(j=0;j<con2;j++) comb[i][j] = - 1;
   }
}

int Search::safe(int v[])
{
    if((v[3]==v[2]) && (v[3]!=v[0]))  // goat eats cabbage
    return 0;
    if((v[3]==v[1]) && (v[3]!=v[0]))  // wolf eats goat
    return 0;
    return 1;   // everything else is safe
}

void Search::cmbgen()
{
    int cmb=16;  // cmb is the number of combinations
    int* v;
```

```
  v = new int[n];  // n is the number of objects
  int y = 0;
  int c;
  for(c=0;c<cmb;c++)
  {
  int x = c;
  int i;
  for(i=0;i<n;i++)
  {
  v[i]= x%2;
  x = (x-v[i])/2;
  }
  if(safe(v) == 1)
  {
  cout <<endl;
  int j;
  for(j=0;j<n;j++)
  {
  comb[j][y] = v[j];
  }
  y += 1;
  }
  }
  delete [] v;
}

void Search::follow()
{
 int v, k;
 for(k=0;k<con1;k++)
 {
 int j;
 for(j=0;j<con1;j++)
 {
 if(comb[0][j] != comb[0][k])
 {
 int t = 0;
 int i;
 for(i=1;i<n;i++)
 {
 if(comb[i][j] != comb[i][k])
 {
 t += 1; v = i;
 }
 }
```

```
  if(((t==1)&&(comb[0][k]==comb[v][k]))||(t==0))
  {
  int x = 0;
  int p;
  do
  {
  x += 1;
  p = comb[x+3][k];
  } while(p!=-1);
  comb[x+3][k] = j;
  }
  }
  }
  }
  comb[5][4] = -1;
}

void Search::sequ()
{
   int k, i, x, z, m, b;
   for(b=0;b<con3;b++) trc[b]=-1;
   int d = 0;
   trc[0] = 0;
   z = 1;
   x = 0;
   do
   {
   i = n;
   m = 0;
   do
   {
   int j;
   for(j=0;j<con3;j++)
   {
   if(comb[i][x]==trc[j]) comb[i][x]=-1;
   }
   k = 0;
   if((comb[i][x] != -1)||(trc[z] != -1))
   {
   if(comb[i][x] !=-1)
   {
   trc[z] = comb[i][x];
   comb[i][x] = -1;
   x = trc[z];
   if(trc[z] == 9) d = con3;
```

```
   }
   z += 1;
   k = 1;
   }
   i += 1;
   m += 1;
   if(m==n) k = 1;
   } while(k!=1);
   d += 1;
   } while(d < con3);

   cout << "The sequence to follow is" << endl << endl;
   cout << "F  W  C  G"<< endl;
   int f;
   for(f=0;f<8;f++)
   {
   int q;
   for(q=0;q<n;q++) cout << comb[q][trc[f]] << "  ";
   cout << "\n";
   }
   cout << endl;
}

void display()
{
    int i, j;
    for(i=0;i<con1;i++)
    {
    cout <<endl;
    for(j=0;j<n;j++) cout << comb[j][i] << "  ";
    }
    cout <<endl;
}

void main()
{
    Search k;
    k.set();
    k.cmbgen();
    cout << "The possible safe combinations is" << endl;
    cout << "F  W  C  G" << endl;
    display();
    k.follow();
    k.sequ();
}
```

11.13 Roman Numbers

All the earliest batteries of number signs in use among priestly astronomers responsible for the custody of the calender disclose a repetition of strokes. Of this there is more than a relic in the familiar *Roman numerals*

$$I, \quad II, \quad III, \quad XX, \quad XXX, \quad CC$$

for 1, 2, 3, 20, 30, 200 etc.. The next step was to group strokes in a regular way and introduce signs for such groups to expedite counting and to economize signs for 5, 50, and 500

$$V, \quad L, \quad D$$

which the Romans took over from the Etruscans as well as 10, 100, 1000 (X, C, M). For example we have

$$146 = CXLVI$$

In the program the function `to_roman` converts an arabic number into a roman number. The function `atoi` converts a character string to an integer. The name stands of *a*lphabetic *to* *i*nteger. The function is prototyped in the `<stdlib.h>` header file. Thus we must include `<stdlib.h>` at the top of any program that uses the `atoi` function. The function `strcat` appends one string to another.

The program `rotoint.cpp` converts a roman number into an arabic number. An example is

$$MCMXCIV = 1994$$

```
// roman.cpp

#include <iostream.h>
#include <string.h>
#include <stdlib.h>

char *roman[] = { "M", "1000", "CM", "900", "D", "500", "CD", "400",
  "C", "100", "XC", "90", "L", "50", "XL", "40",
  "X", "10", "IX", "9", "V", "5", "IV", "4", "I", "1" };

void to_roman(char *buf,int n)
{
   int i, t;
   for(buf[0]='\0',i=0;n>0 && roman[i]; i+=2)
   {
   for(t=atoi(roman[i+1]); n >= t; n -=t)
   strcat(buf,roman[i]);
   }
}

void main()
{
   char* s = new char[20];
   int n = 146;      // number to be converted to roman
   to_roman(s,n);
   cout << " roman number = " << s << "\n";
}
```

```
// rotoint.cpp

#include <iostream.h>
#include <stdio.h>
#include <string.h>      // for strlen
#include <ctype.h>
#include <stdlib.h>

int romantoint(char* roman)
{
   int letters[7] = { 'I', 'V', 'X', 'L', 'C', 'D', 'M' };
   int values[7] = { 1, 5, 10, 50, 100, 500, 1000 };
   const int nvals = 7;
   int lastval = 0;
   int c, i;
```

```
  int n = 0;

  char *s = roman + strlen(roman);
  while(--s >= roman)
  {
     if(islower(*s))
     c = *s - 32;
     else c = *s;

     for(i=0; i < nvals; i++)
     if(c == letters[i]) break;

     if(i == nvals) continue;

     if(values[i] >= lastval)
     n += values[i];
     else
     n -= values[i];
     lastval = values[i];
  }
  return n;
}

int main()
{
  char roman[64];
  for(;;)
  {
  cout << "enter roman numerals: " << endl;
  cout << "type q for quit" << endl;
  cout << "roman = ";

  cin >> roman;

  if(roman[0] == 'q') break;
  printf("%s = %d\n",roman,romantoint(roman));
  }

  return 0;
}
```

11.14 Day of the Week

Given the year, month and day the following program calculates the day of the week. For example, let

```
year = 1994,    month = 4,    day = 29
```

We find that this date is a Friday. We take the following identification

$$\text{Sunday} = 0, \quad \text{Monday} = 1, \quad \text{Tuesday} = 2, \quad \text{Wednesday} = 3$$
$$\text{Thursday} = 4, \quad \text{Friday} = 5, \quad \text{Saturday} = 6$$

The function `weekday` determines the weekdays using the identification given above. For example the battle of Waterloo took place on 1815/6/18. We find that the battle took place on a Sunday.

```
// dayofweek.cpp

#include <iostream.h>

int weekday(int y,int m,int d)
{
   static int t[] = { 0, 3, 2, 5, 0, 3, 5, 1, 4, 6, 2, 4 };
   y -= m < 3;
   return (y + y/4 - y/100 + y/400 + t[m-1] + d)%7;
}

char *days[] = { "Sunday", "Monday", "Tuesday", "Wednesdy",
"Thursday", "Friday", "Saturday" };

void main()
{
    cout << endl;
    int year = 1994;
    int month = 4;
    int day = 29;
    int result = weekday(year,month, day);
    cout << "number of day = " << result << endl;
    cout << "week day = " << days[result] << endl;
}
```

11.15 Gregorian Calender

A calendar is a means of grouping days in ways convenient for regulating civil life and religious observances and for historical and scientific purposes. The word is derived from the Latin calendarium, meaning interest register, or account book, itself a derivation from calendae (or kalendae), the first day of the Roman month, the day on which future market days, feast and other occasions were proclaimed.

The Julian calendar year of 365,25 days was too long, since the correct value for the tropical year is 365,242199 days. This error of 11 minutes 14 seconds per year amounted to almost one and a half days in two centuries, and seven days in 1,000 years. Once again the calendar became increasingly out of phase with the seasons. From time to time, the problem was placed before church councils, but no action was taken because the astronomers who were consulted doubted whether enough percise information was available for a really accurate value of the tropical year to be obtained. By 1545, however, the vernal equinox, which was used in determining Easter, had moved 10 days from its proper date. By the time of Pope Gregory XIII in the late 16th century, the deviation amounted to 10 days; the equinox of 1582 fell on March 11.

The true length of the solar year was by that time known, and Gregory undertook to reform the calendar accordingly. He cut 10 days from the year 1582 so that, starting the following year, the equinox would again fall on March 21. The day after Thursday, October 4, 1582, thus became Friday, October 15. Note that the continuity of the days of the week was maintained. This gap continued to separate the Julian from the Gregorian calendar where the latter was not adopted. Gregory then promulgated an edict establishing that 3 days would be dropped from the calendar every 400 years, thus correcting the discrepancy between the Julian calendar and the solar year. It was decreed that years with a date ending in two zeros would not be leap years, except when the first two digits were divisible by 4. Thus, in the old Julian calendar, 1600, 1700, 1800, 1900, 2000, and so forth would be leap years. In the Gregorian calendar, 1700, 1800, and 1900 were not leap years; but 1600 was, and so will be 2000. This elimination of 3 days every 4 centuries is clear and simple.

There is a remaining inexactness in that the Gregorian year is equal to an average of 365.2425 days and is therefore longer than the solar year by 0.0003 day per year. The excess amounts to 3 days every 10,000 years. However other factors come into play over such long periods. The solar year has been taken here as being exactly 365.2422 days long, but in reality it varies in length (although very slowly). The law governing this variation is imperfectly known. In addition, the rotation of the earth on its axis is subject to variations, some of which cannot be predicted. It would be futile to anticipate the calendar corrections that might be necessary in 3,000 to 4,000 years, when far greater disparities may intervene in the meantime. The Gregorian calendar therefore leaves nothing to be desired in its precision.

We identify

`1 = Monday, 2 = Tuesday, 3 = Wednesday, 4 = Thursday,`

`5 = Friday, 6 = Saturday, 7 = Sunday`

```
// gregor.cpp

#include <iostream.h>
#include <string.h>  // for strcpy
#include <stdlib.h>  // for itoa
#include <time.h>    // for time

     // Declaration of class Date
class Date {
public:
  Date(int Day=1,int Month=1,int Year=2000);
  void today();      // use date from MS-DOS
  int Weekday();     // 1=Mo, 2=Tu, 3=We, ..., 7=Su
  char* String();    // example: "Mo, 31.12.1990"
  int Day();
  int Month();
  int Year();
  void Easter(int Year);
  void operator += (long Days);
  void operator -= (long Days);
  friend long operator - (Date,Date); // Difference in Days
protected:
  int t, m, j;                        // Day, Month, Year
  long Daynumber;
  void calculateDaynumber();
  void calculateTMJ();
};

extern char *MoName[12];
extern char WoDay[7][3];

extern int Monthslength(int Month,int Year=1);
extern int leapyear(int Year);

   // inline functions
```

```
inline int Date::Day()      { return t; }
inline int Date::Month()    { return m; }
inline int Date::Year()     { return j; }

void getDate(int& Day,int& Month,int& Year)
{
  time_t long_time;
  time(&long_time);
  struct tm *tme;
  tme = localtime(&long_time);
  Day = tme -> tm_mday;  Month = tme -> tm_mon;  Year = tme -> tm_year;
}

char *MoName[12] = {
  "January","February","March", "April","May","June",
  "July","August","September","October","November","December"
};

char WoDay[7][3] = { "Mo","Tu","We","Th","Fr","Sa","Su" };

static int MoBegin[12] = {
     0, 31, 61, 92, 122, 153, 184, 214, 245, 275, 306, 377 };

    // Definition of class Date
Date::Date(int Day,int Month,int Year)
{
  t = Day;  m = Month;  j = Year;
  calculateDaynumber();
}

void Date::today()
{
  int t,m,j;
  getDate(t,m,j);
  *this = Date(t,m,j);
}

int Date::Weekday ()
{
  return 1 + (Daynumber+2)%7;
}

char *Date::String()
{
  static char s[15];
```

```
  strcpy(s,WoDay[Weekday()-1]);
  s[2] = ',';
  itoa (t + 100, s+3, 10);
  s[3] = ' ';
  itoa (m + 100, s+6, 10);
  s[6] = s[9] = '.';
  itoa(j,s+10,10);
  return s;
}

void Date::Easter(int Year)
{
  int q = Year/4;
  int a = Year%19;
  int b = (204 - 11*a)%30;
  if(b > 27) b--;
  int c = (Year + q + b - 13)%7;
  int t = 28 + b - c;
  if(t > 31) *this = Date (t-31,4,Year);
  else *this = Date(t,3,Year);
}

void Date::operator += (long Days)
{
  Daynumber += Days;
  calculateTMJ ();
}

void Date::operator -= (long Days)
{
   Daynumber -= Days;
   calculateTMJ();
}

long operator - (Date D1,Date D2)
{
  return D1.Daynumber - D2.Daynumber;
}

static long Marchday(int j)
{
  return 365L*j + j/4 - j/100 + j/400;
}

void Date::calculateDaynumber() {
```

```
  int M = m, J = j;
  if(M > 2) M -= 3;
  else { M +=9; J--; }
  Daynumber = Marchday(J) + MoBegin[M] + t - 1;
}

void Date::calculateTMJ()
{
  int Rest;
  j = 1 + Daynumber/365;
  do {
  j--;
  Rest = Daynumber - Marchday(j);
  }
  while(Rest < 0);
  m = 11;
  while(MoBegin[m] > Rest)
  m--;
  t = Rest + 1 - MoBegin[m];
  if(m > 9) { m -= 9; j++; }
  else m += 3;
}

int Monthslength(int Month,int Year)
{
  static int MoLg[12] = { 31,28,31,30,31,30,31,31,30,31,30,31 };
  if(Month == 2) return leapyear (Year) ? 29 : 28;
  else return MoLg[Month-1];
}

int leapyear(int Year)
{
  return (Year%4 == 0) && ((Year%100 != 0) || (Year%400 == 0));
}

void main()
{
   Date Willi(20,5,1999), now, Fest;
   now.today();
   int j = Willi.Weekday();
   cout << "The weekday is: " << j;
   char *n[] = { "0", "Monday", "Tuesday", "Wednesday", "Thursday",
                 "Friday", "Saturday", "Sunday" };
   cout << " = " << n[j];
}
```

11.16 Records

Given a data file `record.old` the following program `MRecords.cpp` updates the records within this data file and writes to the file `record.new`. First we provide the current date, and then enter a loop that will read a series of customer records from the data file `record.old` and write the corresponding updated records to a new date file called `record.new`. Each pass through the `while` loop will read one record, update it if necessary, and then write the record to `record.new`. All the records will be written to `record.new` whether updated or not. The program will continue until the customer name `END` has been read from the source file. Once this happens, `END` will be written to the data file `record.new`, indicating an end-of-file condition. To leave the `while` loop we use the `break` statement.

The function

```
int strcmpi(const char* string1,const char* string2)
```

compares two strings without regard to case. It returns a value less than, equal to, or greater than 0, depending on whether the string pointed to by `string1` is less than, equal to, or greater than the string pointed to by `string2`. The function `strcmpi` is the same as `stricmp`.

The function

```
int fscanf(FILE* stream,const char* format,[,address,...])
```

scans and formats input from a stream. The function `fscanf` scans a series of input fields, one character at the time, reading from a stream.

The function

```
int fprintf(FILE* stream,const char* format[,argument,...])
```

writes formatted output to a stream.

We also use C++ file manipulation in the program (see chapter 9).

An interactive session would be:

```
Customer Invoice System - Update

Please enter today's date:
Enter month:  3
Enter day:   12
Enter year:  1999

Name: Willi Hans Steeb  Account Number: 2345

New Balance: 1000.00   Current Payment: 50.00
New Balance:  50.00


Name: Nela de Sousa  Account Number: 875

New Balance: 500.00   Current Payment: 250.00
New Balance: 250.00


Name: Olli Xani Moloko  Account Number: 2345

New Balance: 1000.00   Current Payment: 800.00
New Balance:  200.00
```

```
// MRecords.cpp

#include <stdio.h>
#include <iostream.h>
#include <fstream.h>
#include <string.h>

/* Pointers to pre-defined structure FILE */
FILE* fpo;
FILE* fpn;

class Date
{
   public:
     int month;
     int day;
     int year;
};

class Record
{
   public:
     char name[80];
     char street[80];
     char city[80];
     int acct_no;
     char acct_type;
     float oldbalance;
     float newbalance;
     float payment;

     Date lastpayment;
};

// forward declaration
Record readfile(Record customer);
Record update(Record customer,int m,int d,int y);
void writefile(Record customer);

int main()
{

   int flag = 1;
```

```
int m, d, y;

Record customer;

/* open data file */
fpo = fopen("record.old", "r");
fpn = fopen("record.new", "w");

// enter current date
cout << "Customer Invoice System - Update\n\n";
cout << "Please enter today's date: \n";

cout << "Enter month: ";
cin >> m;

cout << "\nEnter day: ";
cin >> d;

cout << "\nEnter year: ";
cin >> y;

while(flag)
{

fscanf(fpo, " %[^\n]", customer.name);
fprintf(fpn, "\n%s\n", customer.name);

if(strcmpi(customer.name, "END") == 0)
break;

customer = readfile(customer);

customer = update(customer,m,d,y);

writefile(customer);
}

fclose(fpo);
fclose(fpn);

char ch;

ifstream f1("record.new");
ofstream f2("record.old");
```

```
   if(!f1) cerr << "cannot open record.new for input";
   if(!f2) cerr << "cannot open record.old for input";

   while((f2 != NULL) && (f1.get(ch) != NULL))
   f2.put(ch);

   f1.close();
   f2.close();

   return 0;
}

Record readfile(Record customer)
{
   fscanf(fpo, " %[^\n]", customer.street);
   fscanf(fpo, " %[^\n]", customer.city);
   fscanf(fpo, " %d", &customer.acct_no);
   fscanf(fpo, " %c", &customer.acct_type);
   fscanf(fpo, " %f", &customer.oldbalance);
   fscanf(fpo, " %f", &customer.newbalance);
   fscanf(fpo, " %f", &customer.payment);
   fscanf(fpo, " %d", &customer.lastpayment.month);
   fscanf(fpo, " %d", &customer.lastpayment.day);
   fscanf(fpo, " %d", &customer.lastpayment.year);

   return customer;
}

Record update(Record customer,int m,int d,int y)
{
   printf("\n\nName: %s", customer.name);
   printf("  Account number: %d\n", customer.acct_no);
   printf("\nNew Balance: %7.2f", customer.newbalance);

   printf("  Current payment: ");
   scanf("%f", &customer.payment);

   if(customer.payment > 0.0)
   {
   customer.lastpayment.month = m;
   customer.lastpayment.day = d;
   customer.lastpayment.year = y;

   customer.oldbalance = customer.newbalance;
   customer.newbalance = customer.newbalance - customer.payment;
```

```
   printf("New balance: %7.2f", customer.newbalance);
   }

   return customer;
}

void writefile(Record customer)
{
   fprintf(fpn,"%s\n", customer.street);
   fprintf(fpn,"%s\n", customer.city);
   fprintf(fpn,"%d\n", customer.acct_no);
   fprintf(fpn,"%c\n", customer.acct_type);
   fprintf(fpn,"%.2f\n", customer.oldbalance);
   fprintf(fpn,"%.2f\n", customer.newbalance);
   fprintf(fpn, "%.2f\n", customer.payment);
   fprintf(fpn, "%d\n", customer.lastpayment.month);
   fprintf(fpn, "%d\n", customer.lastpayment.day);
   fprintf(fpn, "%d\n", customer.lastpayment.year);
}
```

The file `record.old` is given by

```
Willi Hans Steeb
123 Kingsway
Egoli
2345
C
1000.00
100.00
900.00
5
8
1998

Nela de Sousa
23 Main street
Pretoria
875
O
2000.00
500.00
1500.00
5
6
1999

Olli Xani Moloko
17 Church street
CapeTown
1234
C
4000.00
1000.00
3000.00
5
8
1995

END
```

11.17 Finite State Machine

A digital computer may be viewed as a machine which is in a certain internal state at any given moment. The computer reads an input symbol, and then prints an output symbol and changes its state. The output symbol depends solely upon the input symbol and the internal state of the machine, and the internal state of the machine depends solely upon the preceding state of the machine, and the preceding input symbol. The number of states, input symbols and output symbols are assumed to be finite. The definition of a finite state machine is as follows.

A *finite state machine* M consists of five things:

1) A finite set A of inout symbols

2) A finite set S of internal states

3) A finite set Z of output symbols

4) A next-state function f from $S \times A$ into S.

5) An output function g from $S \times A$ into Z.

This machine M is denoted by $M = \langle A, S, Z, f, g \rangle$ when we want to designate its five parts. Sometimes we are also an initial state q_0 in S, and then the machine M is designated by the six-tuple $M = \langle A, S, Z, q_0, f, g \rangle$.

The following coin machine realizes a finite state machine. The machine can be broken by entering a 0.

```
// finite.cpp

#include<iostream.h>

enum cointype { nickel, dime, quarter, slug };
enum statetype { S0, S1, S2, S3, S4, S5 };

cointype coin;
statetype S;

void getcoin(cointype &coin)
{
  int coinval;

  cout<<"\n Deposit coin \n";
  cin >> coinval;
```

```
  while(!((coinval==5)||(coinval==10)||(coinval==25)||(coinval==0)))
  {
  cout << "\nCoin rejected, deposit a valid coin.\n";
  cin >> coinval;
  }
  switch(coinval)
  {
  case 0 : coin = slug;
  break;
  case 5 : coin = nickel;
  break;
  case 10 : coin = dime;
  break;
  case 25 : coin = quarter;
  break;
  }
}

int main()
{
  S = S0;
  getcoin(coin);
  while(coin != slug) {
    switch(S) {

    case S0:
      switch(coin)
      {
      case nickel : S = S1; break;
      case dime : S = S2; break;
      case quarter : S = S5; break;
      }
      break;

    case S1:
      switch(coin)
      {
      case nickel : S = S2; break;
      case dime : S = S3; break;
      case quarter : S = S0; cout<<"\nGUM\n"; break;
      }
      break;

    case S2:
      switch(coin)
```

```
      {
      case nickel : S = S3; break;
      case dime : S = S4; break;
      case quarter : S = S0; cout<<"\nGUM + 5 cents change\n"; break;
      }
      break;

    case S3:
      switch(coin)
      {
      case nickel : S = S4; break;
      case dime : S = S5; break;
      case quarter : S = S0; cout<<"\nGUM + 10 cents change\n"; break;
      }
      break;

    case S4:
      switch(coin)
      {
      case nickel : S = S5; break;
      case dime : S = S0; cout<<"\nGUM\n"; break;
      case quarter : S = S0; cout<<"\nGUM + 15 cents change\n"; break;
      }
      break;

    case S5:
      switch(coin)
      {
      case nickel : S = S0; cout<<"\nGUM\n"; break;
      case dime : S = S0; cout<<"\nGUM + 5 cents change\n"; break;
      case quarter : S = S0; cout<<"\nGUM + 20 cents change\n"; break;
      }
      break;
    }
    getcoin(coin);
  }
  cout<<"\n\nMachine is broken\n\n";

  return 0;
}
```

11.18 Realtor

The class House creates a house with its respective rooms via interaction with the prospective buyer.

```
// mhouse.cpp

#include <iostream.h>

class House
{
   private:
   // declaration of class Room
   // everything that is not a kitchen or a bathroom
      class Room
      {
        private:
           int Doors;
           int Windows;
           int Cupboards;
        public:
           Room()
           {
           Doors = Windows = Cupboards = 0;
           }
           void CreateRoom(int d,int w,int c)  // member function
           {
           Doors = d;
           Windows = w;
           Cupboards = c;

           }
           void ShowR()
           {
           cout << "- " << Doors << " Door(s)" << endl;
           cout<< "- " << Windows << " Window(s)" << endl;
           cout << "- " << Cupboards << " Cupboard(s)" << endl;
           }
        };
        Room* r;
        void GetR(int,char*); // member function: retrieve room details
        void CreateR(int);    // member function: create room

        // declaration of class Bathroom
        // Bathroom inherits basic characteristics from Room
```

```
class Bathroom : public Room
{
   private:
      int Basin;
      int Bath;
      int Shower;
      int Toilet;
   public:
      Bathroom() : Room()  // constructor
      { Basin = Bath = Shower = Toilet = 0; }
      void CreateBathroom(int basin,int bath,int s,
                          int t,int d,int w,int c)
      {  //set bathroom variables
          CreateRoom(d,w,c);
          Basin = basin;
          Bath = bath;
          Shower = s;
          Toilet = t;
      }
      void ShowB()
      {  // display bathroom details
          ShowR();
          cout << "- " << Basin << " Basin(s)" << endl;
          cout << "- " << Bath << " Bath(s)" << endl;
          cout << "- " << Shower << " Shower(s)" << endl;
          cout << "- " << Toilet << " Toilet(s)" << endl;
      }
   };
   Bathroom* b;
   void CreateB(int);  // member function: create Bathroom

   // declaration of class Kitchen
   class Kitchen : public Room
   {  // kitchen inherits basic characteristics from Room
      private:
         int Basin;
         int Stove;
      public:
         Kitchen() : Room()  // constructor
         { Basin = Stove = 0; }
         void CreateKitchen(int b,int s,int d,int w,int c)
         {  // set kitchen variables
            CreateRoom(d,w,c);
            Basin = b;
            Stove = s;
```

```
                }
                void ShowK()
                { // display kitchen details
                   ShowR();
                   cout << "- " << Basin << " Basin(s)" << endl;
                   cout << "- " << Stove << " Stove(s)" << endl;
                }
          };
          Kitchen* k;
          void CreateK(int);  // member function: create Kitchen

          int dnum, wnum, cnum;
          int bnum, rnum, knum;

       public:
          House();  // constructor
          House(int,int,int);
          ~House();  // destructor
          House(const House&);   // copy constructor
          House& operator = (const House&);  // overload operator
          void Show();  // member function

};

 // implementation of House
House::House() // create house
{
   int rn, bn, kn;

   cout << "How many rooms (study, bedrooms, lounge etc.) "
             "do you want: ";
   cin >> rn;
   cout << "How many bathrooms do you want: ";
   cin >> bn;
   cout << "How many kitchens do you want: ";
   cin >> kn;

   CreateR(rn);  // create rooms
   CreateB(bn);  // create bathrooms
   CreateK(kn);  // create kitchens
}

// constructor with no interview with user
House::House(int rn, int bn, int kn)
{
```

```
    bnum = bn;                    // create empty bathroom
    b = new Bathroom[bnum];
    rnum=rn;
    r = new Room[rnum];      // create empty Room
    knum=kn;
    k = new Kitchen[knum];  // create empty kitchen
}

House::~House()    // destructor
{
   delete []k;
   delete []b;
   delete []r;
}

House::House(const House& H)  // copy constructor
{
   int i;
   rnum = H.rnum;
   bnum = H.bnum;
   knum = H.knum;
   dnum = H.dnum;
   wnum = H.wnum;
   cnum = H.cnum;

   r = new Room[rnum];   // copy rooms
   for(i=0; i<rnum; i++)
   {
   r[i] = H.r[i];
   }

   b = new Bathroom[bnum];  // copy bathroom
   for(i=0;i<bnum;i++)
   {
   b[i] = H.b[i];
   }

   k = new Kitchen[knum];  // copy kitchen
   for(i=0;i<knum;i++)
   {
   k[i] = H.k[i];
   }
}

House& House::operator = (const House& H)
```

```
{
   int i;

   if(this != &H)
   {
    delete []k;
    delete []b;
    delete []r;

    rnum = H.rnum;  // assign variables
    bnum = H.bnum;
    knum = H.knum;
    dnum = H.dnum;
    wnum = H.wnum;
    cnum = H.cnum;

    r = new Room[rnum];  // assign rooms
    for(i=0;i<rnum;i++)
    {
    r[i] = H.r[i];
    }

    b = new Bathroom[bnum];  // assign bathroom
    for(i=0;i<bnum;i++)
    {
    b[i] = H.b[i];
    }

    k = new Kitchen[knum];  // assign kitchen
    for(i=0;i<knum;i++)
    {
    k[i] = H.k[i];
    }
    }
    return *this;
}

void House::GetR(int i, char* s)  // get room details
{
   cout << endl << s << " Number " << i << endl;
   cout << "How many windows do you want? " ;
   cin >> wnum;
   cout << "How many doors do you want? " ;
   cin >> dnum;
   cout << "How many cupboards do you want? ";
```

```
   cin >> cnum;
}

void House::CreateR(int rn)  // create rooms
{
   rnum=rn;
   r = new Room[rnum];
   int i;
   for(i=0;i<rnum;i++)
   {
   GetR(i,"Room");
   r[i].CreateRoom(dnum,wnum,cnum);
   }
}

void House::CreateB(int bn)  // create bathrooms
{
   int bsnum, bhnum, snum, tnum;

   bnum = bn;
   b = new Bathroom[bnum];
   int i;
   for(i=0;i<bnum;i++)
   {
   GetR(i,"Bathroom");
   cout << "How many basins do you want? " ;
   cin >> bsnum;
   cout << "How many baths do you want? " ;
   cin >> bhnum;
   cout << "How many showers do you want? ";
   cin >> snum;
   cout << "How many toilets do you want? ";
   cin >> tnum;
   b[i].CreateBathroom(bsnum,bhnum,snum,tnum,dnum,wnum,cnum);
   }
}

void House::CreateK(int kn)  // create kitchen
{
   int bnum, snum;
   knum = kn;
   k = new Kitchen[knum];
   int i;
   for(i=0;i<knum;i++)
   {
```

```
   GetR(i,"Kitchen");
   cout << "How many basins do you want? " ;
   cin >> bnum;
   cout << "How many stove do you want? " ;
   cin >> snum;
   k[i].CreateKitchen(bnum,snum,dnum,wnum,cnum);
   }
}

void House::Show()  // display house details
{
   int i;
   cout << endl << "Your House will have the following rooms: " << endl
   for(i=0;i<rnum;i++)
   {
      cout << "Room " << i << " with: " << endl;
      r[i].ShowR();
   }
   for(i=0;i<bnum;i++)
   {
      cout << "Bathroom " << i << " with: " << endl;
      b[i].ShowB();
   }
   for(i=0;i<knum;i++)
   {
      cout << "Kitchen " << i << " with: " << endl;
      k[i].ShowK();
   }
}

//  an example program that could be used with the class:
void main()
{
   House H;
   cout << endl << "House H" << endl;
   H.Show();
   House N = H;
   cout << endl << "House N" << endl;
   N.Show();
   House D(4,2,1);
   D = H;
   cout << endl << "House D" << endl;
   D.Show();
}
```

Chapter 12

Applications in Statistics

12.1 Introduction

In this chapter we give a collection of programs useful in statistics. Researchers in natural and social science frequently collect measurements on several variables. Statistics is concerned with statistical methods for describing and analyzing these data.

In the program argeha.cpp we evaluate the arithmetic mean, harmonic mean, and geometric mean. Examples are given where we have to use the harmonic mean and geometric mean instead of the arithmetic mean. Mean values and standard deviation are inverstiagted in section 12.3. In section 12.4 a program is developed that finds the histogram for a series of numbers. Data fitting by least squares are done in section 12.5. The mathematical foundation of this method is also described. An extension to higher order polynomials is considered in section 12.6. The exponential function and sine function can also be included for the fitting. Pseudo-random number generators are introduced in section 12.7. The correlation coefficient is introduced and implemented in section 12.8. The autocorrelation function for a data series is discussed in section 12.9. A more sophisticated random number generator is introduced in section 12.10 using the number π.

12.2 Arithmetic, Harmonic and Geometric Mean

Let $a[0]$, $a[1]$, ...,$a[n-1]$ be n positive real numbers. The *arithmetic mean* A is defined by

$$A := \frac{a[0] + a[1] + \ldots + a[n-1]}{n}. \tag{1}$$

The *geometric mean* is defined by

$$G := (a[0] \cdot a[1] \cdot \ldots \cdot a[n-1])^{1/n} \tag{2}$$

The *harmonic mean* H is defined by

$$\frac{1}{H} := \frac{1}{n}\left(\frac{1}{a[0]} + \frac{1}{a[1]} + \ldots + \frac{1}{a[n-1]}\right) \tag{3}$$

This can be extended to the generalized mean

$$M(t) := \left(\frac{1}{n}\sum_{k=0}^{n-1} a[k]^t\right)^{1/t} \tag{4}$$

We find that

$$\lim_{t\to 0} M(t) = G \tag{5}$$

$$M(1) = A \tag{6}$$

$$M(-1) = H. \tag{7}$$

The relation between the arithmetic, geometric and harmonic mean is as follows:

$$A \geq G \geq H$$

with equality if and only if

$$a[0] = a[1] = \ldots = a[n-1].$$

To calculate x^y we use the identity $(x > 0)$

$$x^y \equiv \exp(y \log(x)).$$

Let us now consider two examples.

Example 1. The electricity tariff has increased by 12 per cent, 8 per cent and 16 per cent per annum over a three-year period. To find the average annual increase in tariff, the geometric mean is used. We find

$$G = (1.12 \cdot 1.08 \cdot 1.16)^{1/3} = 1.1195$$

i.e. an 11.95 % annual increase. Assume the initial price per kilowatt hour was \$ 1. Then the actual price after three years would be \$ 1.40313, i.e.

$$1.40313 = 1 \cdot 1.12 \cdot 1.08 \cdot 1.16.$$

The arithmetic mean overstates the average percentage change.

Example 2. A cyclist travels 50 km/h over a 5 km stretch of road, and 30 km/h over another hilly 5 km stretch of road. The average speed over the 10 km distance is given by the harmonic mean

$$\frac{1}{H} = \frac{1}{2}\left(\frac{1h}{30km} + \frac{1h}{50km}\right)$$

The harmonic mean is

$$H = 37.5\frac{km}{h}.$$

Note that the arithmetic mean would have given a value of 40 km/hour which overstates the true average speed.

The following are two programs for the evaluation of the mean values.

```
// argeha.cpp

#include <iostream.h>
#include <math.h>      // for exp(), log()

class MeanVal
{
   private:
      double *data;
      int j, size;
   public:
      MeanVal(int);
      MeanVal(const MeanVal&);
     ~MeanVal();
      double& operator[] (int);
      double arith();
      double geo();
      double har();
};

MeanVal::MeanVal(int n)
{
   data = new double[n];
   size = n;
}

MeanVal::MeanVal(const MeanVal& arg)
{
   j = arg.j;  size = arg.size;
   int n;
   for(n=0; n<size; n++)
   data[n] = arg.data[n];
}

MeanVal::~MeanVal()
{
   delete [] data;
}

double& MeanVal::operator[] (int i)
{
   return data[i];
}
```

```
double MeanVal::arith()
{
   double s1 = 0.0;
   for(j=0;j<size;j++)
   {
   s1 += data[j];
   }
   return s1/size;
}

double MeanVal::geo()
{
   double s2 = 1.0;
   for(j=0;j<size;j++)
   {
   s2 *= data[j];
   }
   return exp(log(s2)/size);
}

double MeanVal::har()
{
   double s3 = 0.0;
   for(j=0;j<size;j++)
   {
   s3 += 1.0/data[j];
   }
   return size/s3;
}

void main()
{
   MeanVal x(4);
   x[0] = 1.0; x[1] = 2.0; x[2] = 5.5; x[3] = 14.3;

   double armean = x.arith();
   cout << "arithmetic mean = " << armean << "\n";

   double geomean = x.geo();
   cout << "geometric mean = " << geomean << "\n";

   double harmean = x.har();
   cout << "harmonic mean = " << harmean << "\n";
}
```

12.3 Mean Value and the Standard Deviation

Consider a sequence of floating point numbers

$$x_0, x_1, \ldots, x_{N-1}. \tag{1}$$

The *mean value* is defined as

$$\bar{x} := \frac{1}{N} \sum_{i=0}^{N-1} x_i. \tag{2}$$

The *deviation* d_i about the mean value is given by

$$d_i := (x_i - \bar{x}), \qquad i = 0, 1, \ldots, N-1 \tag{3}$$

The *standard deviation* is

$$s := \sqrt{(d_0^2 + d_1^2 + \cdots + d_{N-1}^2)/N} \tag{4}$$

In the program `mean.cpp` we calculate the mean value of a given data set and then the standard deviation. The function `sum()` calculates the sum

$$\sum_{i=0}^{N-1} x_i$$

and the function `deviation()` calculates the standard deviation s.

In the program `average.cpp` the number of data points and the data are read in. The memory is dynamically allocated using the operator `new`.

```
// mean.cpp

#include <iostream.h>    // for cout
#include <math.h>        // for sqrt

double sum(double x[],int n);
double deviation(double x[],int n,double mean);

int main()
{
   const int N = 10;   // number of data points

   double x[N];
   double mean;
   x[0] = 115.0; x[1] = 116.0; x[2] = 128.0; x[3] = 132.0;
   x[4] = 135.0; x[5] = 138.0; x[6] = 140.0; x[7] = 150.0;
   x[8] = 150.0; x[9] = 156.0;
```

```
  mean = sum(x,N)/N;
  cout << " mean = " << mean << "\n";
  double dev = deviation(x,N,mean);
  cout << " deviation = " << dev << "\n";

  return 0;
} // end main

double sum(double x[],int n)
{
  int i;
  double sum = 0.0;

  for(i=0;i<n;i++)
  {
  sum += x[i];  // can also be written as sum = sum + x[i]
  }
  return sum;
}

double deviation(double x[],int n,double mean)
{
  int i;
  double temp = 0;
  double d[N];

  for(i=0;i<n;i++)
  {
  d[i] = x[i] - mean;
  temp += d[i]*d[i];
  }
  double s = sqrt(temp/n);
  return s;
}
```

In the following program the memory for the vector a is dynamically allocated.

```
// average.cpp

#include <iostream.h>
#include <math.h>

void sum(double* a, double *sumx, double *sumx2, int n)
{
   int i;
   for(i=0;i<n;i++)
   {
   *sumx = *sumx + a[i];
   *sumx2 = *sumx2 + a[i]*a[i];
   }
}

int main()
{
   cout << endl << endl;
   double *a = NULL;
   int n;
   cout << "number of data: ";
   cin >> n;
   cout << endl;
   a = new double[n];
   int j;
   for(j=0; j<n; j++)
   {
   cout << "a[" << j << "] = ";
   cin >> a[j];
   }
   double sumx = 0;  double sumx2 = 0;
   sum(a,&sumx,&sumx2,n);
   double av = (1.0/n)*sumx;
   cout << "\n average value = " << av;
   double var = (1.0/(n-1.0)*(sumx2 - (1.0/n)*sumx*sumx));
   cout << "\n variance = " << var;
   cout << "\n standard deviation = " << sqrt(var);

   return 0;
}
```

12.4 Histogram for Series of Numbers

Given eighteen school children in standard 8. The measuring of the height of the school children in cm gives the following values:

```
d[0] = 151.1,   d[1] = 157.3,   d[2] = 163.2,

d[3] = 162.1,   d[4] = 164.9,   d[5] = 168.3,

d[6] = 167.3,   d[7] = 145.8,   d[8] = 164.9,

d[9] = 148.1,   d[10] = 156.6,  d[11] = 162.7,

d[12] = 165.3,  d[13] = 155.5,  d[14] = 160.1,

d[15] = 158.3,  d[16] = 161.4,  d[17] = 148.3
```

Since the smallest child is 145.8 cm and the tallest is 168.3 cm we consider the interval [145,170]. We divide the interval [145,170] into 5 bins and ask how many pupils are in the intervals

```
[145,150),  [150,155),  [155,160),  [160,165),  [165,170]
```

This means we find the frequency distribution (histogram).

```
// histon.cpp

#include <iostream.h>
#include <math.h>

int main()
{
   const int ND = 18;   // number of data
   const int NB = 5;    // number of bins

   float data[ND];
   data[0] = 151.1; data[1] = 157.3; data[2] = 163.2;
   data[3] = 162.1; data[4] = 164.9; data[5] = 168.3;
   data[6] = 167.3; data[7] = 145.8; data[8] = 164.9;
   data[9] = 148.1; data[10] = 156.6; data[11] = 162.7;
   data[12] = 165.3; data[13] = 155.5; data[14] = 160.1;
   data[15] = 158.3; data[16] = 161.4; data[17] = 148.3;

   int hist[NB];

   int i;
   for(i=0;i < NB; i++)
   {
   hist[i] = 0;
   }
   float Xmax = 170.0;  float Xmin = 145.0;
   float Diff = Xmax - Xmin;

   for(i=0;i<ND;i++)
   {
   ++hist[(int) floor(NB/Diff*(data[i]-Xmin))];
   }
   for(i=0;i<NB;i++)
   cout << "hist[" << i << "] = " << hist[i] << "\n";

   return 0;
}
```

12.5 Data Fitting by Least Squares

In data fitting its is not uncommon to use a very simple approximation curve, such as a straight line or parabola, and indeed there are a number of situations in which this is reasonable. Let us consider a straight line, for which explicit results may be calculated easily.

Assume that the independent variable x is exact, while the dependent variable $y = f(x)$ has random errors occuring in a normal distribution. Suppose that there are N data points (x_i, y_i) $(i = 1, \ldots, N)$, and that values of a and b are required to define a straight line approximation

$$y \simeq f^*(x) = a + bx \tag{1}$$

where a is the intercept with the y axis, and b is the gradient. Adopting the *least square norm*, namely

$$s := \|f - f^*\| = \left(\frac{1}{N}\sum_{i=1}^{N}(f(x_i) - f^*(x_i))^2\right)^{1/2} = \left(\frac{1}{N}\sum_{i=1}^{N}(y_i - a - bx_i)^2\right)^{1/2} \tag{2}$$

we seek the best approximation of form (1) which minimises s. Obviously the minimum of s coincides with the minimum of Ns^2, that is of

$$E = \sum_{i=1}^{N}(y_i - a - bx_i)^2. \tag{3}$$

The sum has an extreme value when

$$\frac{\partial E}{\partial a} = 0 \quad \text{and} \quad \frac{\partial E}{\partial b} = 0. \tag{4}$$

Since

$$\frac{\partial E}{\partial a} = -2\sum_{i=1}^{N}(y_i - a - bx_i) = 0 \tag{5}$$

it follows that

$$aN + b\sum_{i=1}^{N} x_i = \sum_{i=1}^{N} y_i. \tag{6}$$

From

$$\frac{\partial E}{\partial b} = -2\sum_{i=1}^{N} x_i(y_i - a - bx_i) = 0 \tag{7}$$

we obtain

$$a\sum_{i=1}^{N} x_i + b\sum_{i=1}^{N} x_i^2 = \sum_{i=1}^{N} x_i y_i. \tag{8}$$

Solving (6) and (8) we obtain for a and b

$$b = \frac{N\sum_{i=1}^{N} x_i y_i - \sum_{i=1}^{N} x_i \sum_{i=1}^{N} y_i}{N\sum_{i=1}^{N} x_i^2 - (\sum_{i=1}^{N} x_i)^2}, \qquad a = \frac{\sum_{i=1}^{N} y_i - b\sum_{i=1}^{N} x_i}{N}. \tag{9}$$

The line of best fit $y = a + bx$ always passes through the centre of gravity of the data points, since

$$\bar{x} := \frac{1}{N}\sum_{i=1}^{N} x_i, \qquad \bar{y} := \frac{1}{N}\sum_{i=1}^{N} y_i, \quad \text{and} \quad a = \bar{y} - b\bar{x}. \tag{10}$$

The root mean square error (1) is an approximation of the standard deviation of the fit defined by

$$\sigma := \left(\frac{1}{N}\sum_{i=1}^{N}(y_i^T - a - bx_i)^2\right)^{1/2} \tag{11}$$

where y_i^T is the true value at x_i of the function underlying the experimental data. However, statistical theory shows that a more realistic estimate of (11) (in which y_i^T is unknown) is given by the estimated standard deviation

$$s_2 = \left(\frac{1}{N-2}\sum_{i=1}^{N}(y_i - a - bx_i)^2\right)^{1/2} \tag{12}$$

The denominator $N - 2$ in the square root, which replaces the denominator N in the error (1), measures the number of degrees of freedom in the data (i.e. data less 2 parameters).

The program `least1.cpp` calculates the quantities a and b. They are defined in the main function. The N ($N = 7$) data points are given as a one-dimensional array. When an array is passed as argument to a function, it is not actually the array that is passed, but only its address. The array itself does not move. In C and C++ there is a strong relationship between pointers and arrays. Any operation that can be achieved by array subscripting can also be done with pointers. A reference to

```
a[i]
```

can also be written as

```
*(a+i)
```

In evaluating `a[i]`, C and C++ converts it to `*(a+i)` immediately. The two forms are equivalent.

In the program `least2.cpp` the memory is dynamically allocated using the operator `new`.

```
// least1.cpp

#include <iostream.h>   // for cout

void least_square(double x[],double y[],int n, double* a, double* b);

void main()
{
   const int N = 7;                    // number of data points
   double x[N], y[N];                  // data points
   x[0] = 2.0; y[0] = 0.7;
   x[1] = 3.0; y[1] = 0.8;
   x[2] = 4.0; y[2] = 1.1;
   x[3] = 5.0; y[3] = 1.2;
   x[4] = 6.0; y[4] = 1.6;
   x[5] = 7.0; y[5] = 1.7;
   x[6] = 8.0; y[6] = 2.0;
   double a = 0.0;
   double b = 0.0;

   least_square(x,y,N,&a,&b);        // call of function least_square

   cout << " a = " << a << "  " << " b = " << b;
} // end main

void least_square(double x[],double y[],int n,double* a,double* b)
{
  int i;
  double sum_x = 0.0, sum_y = 0.0, sum_xy = 0.0, sum_x2 = 0.0;
  for(i=0;i<n;i++)
  {
  sum_x += x[i];  // can also be written as sum_x = sum_x + x[i]
  sum_y += y[i];
  sum_xy += x[i]*y[i];
  sum_x2 += x[i]*x[i];
  }
  *b = (sum_x*sum_y-n*sum_xy)/(sum_x*sum_x-n*sum_x2);
  *a = (sum_y-(*b)*sum_x)/n;
}
```

The output is

$$a = 0.192857, \qquad b = 0.221429$$

```
// least2.cpp

#include <iostream.h>

void least_square(double* ,double* ,int n, double*, double*);

void main()
{
   cout << endl << endl;
   double* x = NULL; double* y = NULL;
   x = new double [n];   y = new double [n];
   int n;
   cout << "number of data: ";
   cin >> n;
   int j;
   for(j=0; j<n; j++)
   {
   cout << "x[" << j << "] = ";
   cin >> x[j];
   cout << "y[" << j << "] = ";
   cin >> y[j]; cout << endl;
   }
   double a = 0.0;   double b = 0.0;
   least_square(x,y,n,&a,&b);

   cout << " a = " << a << "  " << " b = " << b;
} // end main

void least_square(double *x,double *y,int n,double* a, double* b)
{
  int i;
  double sum_x = 0.0, sum_y = 0.0, sum_xy = 0.0, sum_x2 = 0.0;
  for(i=0;i<n;i++)
  {
  sum_x += *(x+i);
  sum_y += *(y+i);
  sum_xy += (*(x+i))*(*(y+i));
  sum_x2 += (*(x+i))*(*(x+i));
  }

  *b = (sum_x*sum_y-n*sum_xy)/(sum_x*sum_x-n*sum_x2);
  *a = (sum_y-(*b)*sum_x)/n;
}
```

12.6 Curve Fitting

Line regression as explained in the previous section works well if the measured data are intrinsically linear or if the range of the abscissae is small. For other cases, however, better results may be obtained by fitting a higher-order polynomial to the data set.

The principle of least squares can be extended to fit a polynomial of any order to measured data. An N-th order polynomial is first written as

$$g(x) = a_0 + a_1 x + a_2 x^2 + \cdots + a_N x^N \tag{1}$$

where $a, x \in \mathbf{R}$. The deviation of the curve from each data point is

$$r_i := y_i - g(x_i), \qquad i = 1, 2, \ldots, L \tag{2}$$

where L is the number of data points. The square total of deviations R is as follows

$$R := \sum_{i=1}^{L} (r_i)^2 . \tag{3}$$

We set the partial derivatives of R with respect to the coefficients of the polynomial to zero to minimize R

$$\frac{\partial R}{\partial a_k} = 0, \qquad k = 0, 1, 2, \ldots, N . \tag{4}$$

It follows that

$$\sum_{n=0}^{N} \left[\sum_{i=1}^{L} x_i^{n+k} \right] a_n = \sum_{i=1}^{L} x_i^k y_i \quad \text{for} \quad k = 0, 1, 2, \ldots, N . \tag{5}$$

Using matrix notation this system of linear equations can be written as

$$\begin{pmatrix} L & \sum_{i=1}^{N} x_i & \sum_{i=1}^{N} x_i^2 & \cdots & \sum_{i=1}^{N} x_i^N \\ \sum_{i=1}^{N} x_i & \sum_{i=1}^{N} x_i^2 & \sum_{i=1}^{N} x_i^3 & \cdots & \sum_{i=1}^{N} x_i^{N+1} \\ & & & \cdots & \\ \sum_{i=1}^{N} x_i^N & \sum_{i=1} x_i^{N+1} & \sum_{i=1}^{N} x_i^{N+2} & \cdots & \sum_{i=1}^{N} x_i^{2N} \end{pmatrix} \begin{pmatrix} a_0 \\ a_1 \\ .. \\ .. \\ a_N \end{pmatrix} = \begin{pmatrix} \sum_{i=1}^{N} y_1 \\ \sum_{i=1}^{N} x_i y_1 \\ \\ \sum_{i=1}^{N} x_i^N y_i \end{pmatrix}$$

The coefficients, a_n, where $n = 0, 1, 2, \ldots, N$, are determined by solving (5) simultaneously using Gauss elimination.

Linear equations that arise in curve fitting are often ill-conditioned when the coefficients of the linear equations become a mix of very small and very large numbers. The spread becomes intensified as the range of x values of data points and the order of the polynomial both increase. Therefore, it is desirable to use double precision in solving the linear equations.

To summarize. Polynomial fitting is an extension of line fitting, and is based on the least squares method. The number of points fitted is generally much greater than the order of the polynomial. The linear equation associated with polynomial fitting is often highly vulnerable to round-off errors. The use of double precision is recommended.

The `goto` statement is used in the program. The format is simple: `goto label`, where `label` is some identifier, associated with a given statement.

```
// CF.CPP
// Curve Fitting with Polynomial Fitting or
// Linear Combination by Least Square

#include <stdio.h>
#include <iostream.h>
#include <stdlib.h>
#include <math.h>       // for fabs, pow, sin, exp

//  x[j], y[j] : data table
//  a[i][j] : matrix
//  a[i][m+1] : coefficients of the curve
//  no_of_data : number of data points

void gauss(int n,double a[][10]) // Gauss elimination
{
   int i, ipv, j, jc, jr, k, kc, nv;
   double r, tm, va;
   for(i=1; i <= (n-1); i++ ) {
   ipv = i;
   for(j=i+1; j <= n; j++)
   if(fabs(a[ipv][i]) < fabs( a[j][i]))  ipv = j;
        if(ipv != i){
           for(jc=1; jc <= (n + 1); jc++ ){
           tm = a[i][jc]; a[i][jc] = a[ipv][jc];  a[ipv][jc] = tm;
 }
      }
      for(jr=i+1; jr <= n; jr++)
      {
      if(a[jr][i] != 0)
      {
      if(a[i][i] == 0.0) {
      cout << "Matrix is singular!\n";
      exit(0);
      }
      r = a[jr][i]/a[i][i];
      for(kc=i+1; kc <= (n+1); kc++)
      a[jr][kc] = a[jr][kc] - r*a[i][kc];
      }
      }
   }
   if(a[n][n] == 0.0)
   {
      cout << "Matrix is singular!\n";
      exit(0);
```

```
   a[n][n + 1] = a[n][n + 1]/a[n][n];
   for(nv=n-1; nv >= 1; nv--) {
   va = a[nv][n + 1];
   for(k=nv+1;k <= n; k++)
   va = va - a[nv][k]*a[k][n+1];
   a[nv][n+1] = va/a[nv][nv];
   }
   return;
}

double fun(int k,double x)
{
   double f;
   if(k==1)  f = 1.0;     // First function
   if(k==2)  f = x;       // Second function
   if(k==3)  f = sin(x); // Third function
   if(k==4)  f = exp(x); // Fourth function
   return f;
}

void main()
{
   int i, j, jex, jj, k, lp, m, n, nord;
   double fk, gg[100], yy, fljj;
   double x[100], y[100];
   double a[10][10], power;
   static no_of_data = 6;
   // data
   static double x_ini[] = { 0.1, 0.4, 0.5, 0.7, 0.7, 0.9 };
   static double y_ini[] = { 0.61, 0.92, 0.99, 1.52, 1.47, 2.03 };

   L_10:     // label for goto
   for(j=1; j <= no_of_data; j++)
   {
   x[j] = x_ini[j-1];
   y[j] = y_ini[j-1];
   }
   cout << "Curve Fitting by Least Square \n";
   cout << "Type 0 for polynomial fitting\n";
   cout << " 1 for linear combination\n";
   cin >> lp;
   if(lp == 0) {  // Polynomial fitting
     cout << " Enter order of polynomial\n";  cin >> nord;
     m = nord + 1;
```

```
    for(k=1; k<=m; k++)
    {
    for(j=1; j<=m+1; j++) a[k][j] = 0.0;
    }
    for(k=1; k <= m; k++)
    {
    for(i=1; i <= no_of_data; i++)
    {
    for(j=1; j <= m; j++)
    {
    jj = k - 1 + j - 1;
    a[k][j] = a[k][j] + pow(x[i],jj);
    }
    yy = pow(x[i],k-1);
    a[k][m+1] = a[k][m+1] + y[i]*yy;
    }
   }
   }
   else {
   cout << " Enter the number of functions in linear combination\n";
   cout << " Four functions are available\n";
   cin >> m;
   for(k=1; k<=m; k++)
   {
   for(j=1; j <= m + 1; j++ )  a[k][j] = 0.0;
   }
   for(k=1; k <= m; k++ ){
     for(i=1; i <= no_of_data; i++)
     {
     fk = fun(k,x[i]);
     for(j=1; j<= m; j++)
     {
     a[k][j]=a[k][j]+fk*fun(j,x[i]);
     }
     a[k][m+1] = a[k][m+1] + y[i]*fk;
     }
  }
  }
  for(i=1; i <= m; i++)
  {
  for(j=1; j <= (m + 1); j++)
  {
  printf(" %11.4e ", a[i][j]);
  }
  cout << "\n";
```

```
  }
  n = m;
  gauss(n,a);
  cout << "Power coefficients determined \n";
  cout << "--------------------------------------\n";
  if(lp == 1) {
  cout << "     Function       Coefficient       \n";
  }
  else { cout << "     Power            Coefficient\n";
  }
  cout << "--------------------------------------\n";
  for(i=1; i <= m; i++) {
     if(lp == 1 ) {
     printf(  "    %3d                   %12.5e \n", i, a[i][m+1]);
     }
     else {
     printf(  "    %3d                   %12.5e \n", i - 1, a[i][m+1]);
     }
  }
  cout << "--------------------------------------\n";
  for(i=1; i <= no_of_data; i++) {
     gg[i] = 0.0;
     for (k=1;k<=m; k++)  {
if(lp == 1) {
        gg[i] = gg[i] + a[k][m+1]*fun(k,x[i]);
        }
        else {
        gg[i] = gg[i] + a[k][m+1]*pow(x[i],k-1);
        }
   }
   }
   cout <<  "Error evaluation\n";
   cout << "-----------------------------------------------------\n";
   printf( "  i     x(i)     y(i)     curve fitted    Derivation\n");
   cout << "-----------------------------------------------------\n";
   for(i=1; i <= no_of_data; i++ ) {
      printf("%4d   %11.4e %11.4e %11.4e %11.4e \n",
             i,x[i],y[i],gg[i],y[i]-gg[i]);
   }
   cout << "-----------------------------------------------------\n";
   cout << "\n Enter 0 to quit, 1 to continue.\n";
   cin >> k;
   if(k != 0) goto L_10;
   exit(0);
}
```

12.7 Pseudo-Random Number Generator

The following program generates *pseudo random number* using the *linear congruential algorithm.* Starting with a seed x_0, a recursive equation is used to obtain the successive random numbers. The parameters used in the one-dimensional difference equation

$$x_{n+1} = (ax_n + b) \mod m$$

are the constants b (a non-negative integer), m, a (positive integers) and $n = 0, 1, 2, \ldots$. In program `rand1.cpp` we use the following parameters

$$a = 3125, \qquad b = 47, \qquad m = 2048\,.$$

Thus we have a period of 2048. Let $x_0 = 7$. Then we find the pseudo-random number sequence

$$x_1 = 1442, \quad x_2 = 697, \quad x_3 = 1148, \ldots$$

The operator `%` is the modulus operator. For example we have $7\%3 = 1$ and $17\%7 = 3$. In program `rand2.cpp` we select parameters with a higher period. The sequence is not always very "pseudo-random" for all choices of a, b, m and x_0.

```
// rand1.cpp
#include <iostream.h>
#include <stdlib.h>

   // Definition of class RandNum
class RandNum
{
   private:
      unsigned int x0;
      static unsigned int a, b, m;
   public:
      RandNum(unsigned int);          // constructor one argument
      virtual unsigned int map();  // member function
};

// Implementation of class RandNum
RandNum::RandNum(unsigned int i) { x0 = i; }

unsigned int RandNum::map()
{
   unsigned int x, k;
   x = (x0*a + b)%m;
   return x;
}
```

```
 // Definition of parameters
unsigned int RandNum::a = 3125; unsigned int RandNum::b = 47;
unsigned int RandNum::m = 2048;

void main()
{
   int x0, N;  unsigned int M;
   cout << "Enter the initial value x0 = ";  cin >> x0;
   cout << "Number of random numbers needed = ";  cin >> N;
   cout << "\n";
   int i;
   for(i=0;i<N;i++)
   {
   RandNum R(x0);  x0 = R.map();
   cout << "random number [" << i << "] = " << x0 << "\n";
   };
   cout << "\n";
}

// rand2.cpp
#include <iostream.h>

double rand(double seed)
{
   long unsigned M = 2147483648;  long unsigned A = 843314861;
   long unsigned C = 453816693;
   long S;
   S = (long) (seed*M);
   S = S*A + C;
   if(S < 0)  S = S + M;
   return ((double) S)/((double) M);
}

void main()
{
   int seed = 5;
   double x = rand(seed);
   int j;
   for(j=0; j<10; j++)
   {
   x = rand(x);
   cout << "x = " << x << endl;
   }
}
```

12.8 Correlation Coefficient

Consider two N-dimensional vectors

$$\mathbf{x} = (x_1, x_2, \ldots, x_N)$$

$$\mathbf{y} = (y_1, y_2, \ldots, y_N).$$

The *linear correlation coefficient* r of the two vectors is defined by

$$r := \frac{N\sum_{i=1}^{N} x_i y_i - \sum_{i=1}^{N} x_i \sum_{i=1}^{N} y_i}{\sqrt{(N\sum_{i=1}^{N} x_i^2 - (\sum_{i=1}^{N} x_i)^2)(N\sum_{i=1}^{N} y_i^2 - (\sum_{i=1}^{N} y_i)^2)}}$$

and measures the strength of the linear relationship between $\mathbf{x}$ and $\mathbf{y}$. If the $\mathbf{x}$ and $\mathbf{y}$ values are related by

$$\mathbf{y} = c\mathbf{x}$$

where c is any positive constant we find

$$r = 1.$$

If the $\mathbf{x}$ and $\mathbf{y}$ vectors are related by

$$\mathbf{y} = -c\mathbf{x}$$

where c is any positive constant we find

$$r = -1.$$

Completely uncorrelated data will give

$$r = 0.$$

The following program `correl.cpp` calculates the correlation coefficient for a number of data points given by the user. The program first ask the user to enter the number of data. Then the data itself are entered by the user. Thus the memory for the vector has to be dynamically allocated.

```
// correl.cpp

#include <iostream.h>
#include <math.h>       // for sqrt


void sums(double* x,double* y,int n,double* a,double* b,double* c,
          double* d,double* e);

void main()
{
   int n, j;
   double* x = NULL;  double* y = NULL;
   x = new double[n];   y = new double[n];
   cout << "number of data points: ";   cin >> n;
   for(j=0; j<n; j++)
   {
   cout << "x[" << j << "] = ";   cin >> x[j];
   cout << "y[" << j << "] = ";   cin >> y[j];
   }
   double a, b, c, d, e;
   a = 0.0;  b = 0.0;  c = 0.0;  d = 0.0;  e = 0.0;
   sums(x,y,n,&a,&b,&c,&d,&e);
   double r = (n*c-a*b)/sqrt((n*d-a*a)*(n*e-b*b));
   cout << " r = " << r << "\n";
} // end main

void sums(double* x,double* y,int n,double* a,double* b,
  double* c,double* d,double* e)
{
  int i;
  double sum_x = 0.0, sum_y = 0.0;
  double sum_xy = 0.0, sum_x2 = 0.0, sum_y2 = 0.0;
  for(i=0;i<n;i++)  {
  sum_x += x[i];  // can also be written as sum_x = sum_x + x[i]
  sum_y += y[i];
  sum_xy += x[i]*y[i];
  sum_x2 += x[i]*x[i];
  sum_y2 += y[i]*y[i];
  }
  *a = sum_x; *b = sum_y;
  *c = sum_xy; *d = sum_x2; *e = sum_y2;
}
```

12.9 Autocorrelation Function

Consider the time sequence

$$x_0, \quad x_1, \quad x_2, \quad x_3, \quad x_4, \ldots$$

The *time average* is defined as

$$\langle x_t \rangle := \lim_{T \to \infty} \frac{1}{T} \sum_{t=0}^{T} x_t \, .$$

Obviously, $\langle x_t \rangle$ depends on the initial value x_0. The *autocorrelation function* is defined as

$$C_{xx}(\tau) := \lim_{T \to \infty} \frac{1}{T} \sum_{t=0}^{T} (x_t - \langle x_t \rangle)(x_{t+\tau} - \langle x_t \rangle) \, .$$

In the program we calculate the time average and the autocorrelation function for the recursion relation

$$x_{t+1} = 4x_t(1 - x_t), \qquad t = 0, 1, 2, \ldots$$

The exact value for the time average is $\langle x_t \rangle = 1/2$. The exact solution for the autocorrelation function is

$$C_{xx}(\tau) = \begin{cases} \frac{1}{8} & \text{for} \quad \tau = 0 \\ 0 & \text{otherwise} \end{cases}$$

This agrees quite well with the output of the program, where

$$\langle x_t \rangle = 0.506$$

and

$$C(0) = 0.123, \quad C(1) = -0.01, \quad C(2) = 0.01.$$

The program listing is:

```
//autocorr.cpp

#include <iostream.h>

class Data
{
   private:
      double* vec;
      int size;
   public:
      Data();
      Data(int);
      double& operator[](int);
      double mean();
      double autocorr(int);
};

Data::Data() { }

Data::Data(int n)
{
   vec = new double [n];
   size = n;
}

double& Data::operator [] (int n)
{
   return vec[n];
}

double Data::mean()
{
  double tem = 0.0;
  int j;
  for(j=0;j<size;j++)
  {
  tem+=vec[j];
  }
  return(tem/size);
}

double Data::autocorr(int n)
{
```

```
  double tem=0.0;
  int j;
  for(j=0;j<size-n;j++)
  {
  tem += (vec[j]-this->mean())*(vec[j+n]-this->mean());
  }
  return(tem/size);
  }

int main()
{
  int N = 2000;

  Data x(N);

  x[0] = 0.618;   // initial value

  int j;
  for(j=1;j<N;j++)
  {
  x[j] = 4.0*x[j-1]*(1.0-x[j-1]);
  }
  cout <<" mean = "<< x.mean() << "\n";

  int k;
  for(k=0;k<4;k++)
  cout  <<" corr(" << k << ") = " << x.autocorr(k) <<"\n";

  return 0;
}
```

12.10 Random Number Generator and π

The explosive growth in the use of Monte Carlo simulations in diverse areas of science has prompted extensive investigations on random number generators. It has long been known that poor quality random number generators can lead to systematic errors in Monte Carlo simulations. As finite-state algorithm standard pseudo-random number generators are, of course, ultimately periodic. This early problems with congruential generators (see section 12.7) led to the development of improved methods for producing random numbers. It can be shown that $2 \cdot 10^7$ digits of the number π are the best known random sequence.

Since

$$\pi = 3.14159265358979323846264 33.....$$

we could form the random sequence

$$0.141, \quad 0.592, \quad 0.653, \ldots$$

in the interval $[0, 1]$. Here we split the number π into slices of three digits. On the other hand we could also form a random sequence of integers

$$1, \quad 4, \quad 1, \quad ,5, \quad 9, \ldots$$

and so on.

The following two programs calculate the number π up to a given number of digits.

```
// pi1.cpp

#include <stdio.h>
#include <stdlib.h>

long kf, ks, *mf, *ms;
long cnt, n, temp, nd;
long i;
long col, col1;
long loc, stor[21];
FILE *fp;

void shift(long *l1,long *l2,long lp,long lmod)
{
     long k;
     k = ((*l2)>0 ? (*l2)/lmod : -(-(*l2)/lmod)-1);
     *l2 -= k*lmod;
```

```
    *l1 += k*lp;
}

void yprint(long m)
{
    if(cnt < n)
    {
    if(++col==11)
    {
    col = 1;
    if(++col1==6)
    {
    col1 = 0;
    fprintf(fp,"\n");
    printf("\n");
    fprintf(fp,"%4ld",m%10);
    printf("%4ld",m%10);
    }
    else {
    fprintf(fp,"%3ld",m%10);
    printf("%3ld",m%10);
    }
    }
    else {
    fprintf(fp,"%ld",m);
    printf("%ld",m);
    }
    cnt++;
    }
}

void xprint(long m)
{
    long ii,wk,wk1;
    if(m < 8)
    {
    for(ii=1; ii<=loc;)
    yprint(stor[ii++]);
    loc = 0;
    }
    else
    if(m > 9)
    {
    wk = m/10;
    m %= 10;
```

```
    for(wk1=loc; wk1>=1;wk1--)
    {
    wk += stor[wk1];
    stor[wk1] = wk % 10;
    wk /= 10;
    }
    }
    stor[++loc] = m;
}

void main(int argc,char *argv[])
{
    int i=0;

    stor[i++] = 0;
    if(argc < 2)
    {
    fprintf(stderr,"Format is:\n\tlongpi <# of places>\n");
    exit(0);
    }

    n = atoi(argv[1]);

    mf = (long *)calloc((int)n+3, sizeof(long));
    if(mf==0)
    {
    fprintf(stderr,"Memory allocation failure [mf]\n");
    exit(0);
    }

    ms = (long *)calloc((int)n+3, sizeof(long));
    if(ms==0)
    {
    fprintf(stderr,"Memory allocation failure [ms]\n");
    exit(0);
    }

    printf("Memory allocated.\n");

    fp = fopen("pi.out","w");
    fprintf(fp,"\nThe following is an approximation of PI to %ld \
       digits\n",(long)n);
    printf("\nThe following is an approximation of PI to %ld \
       digits\n",(long)n);
```

```
    cnt = 0;
    kf = 25;
    ks = 57121L;
    mf[1] = 1;
    for(i=2; i<=n; i+=2)
    {
    mf[i] = -16;
    mf[i+1] = 16;
    }

    for(i=1; i<=n; i+=2)
    {
    ms[i] = -4;
    ms[i+1] = 4;
    }
    fprintf(fp,"\n 3.");
    printf("\n 3.");

    while(cnt < n)
    {
    for(i=0; ++i<=n-cnt;)
    {
    mf[i] *= 10;
    ms[i] *= 10;
    }
    for(i=n-cnt+1; --i>=2;)
    {
    temp = 2*i-1;
    shift(&mf[i-1],&mf[i],temp-2,temp*kf);
    shift(&ms[i-1],&ms[i],temp-2,temp*ks);
    }
    nd = 0;
    shift(&nd,&mf[1],1L,5L);
    shift(&nd,&ms[1],1L,239L);
    xprint(nd);
    }
    printf("\n\nCalculations Completed!\n");
}
```

```
// pi2.cpp

#include <stdio.h>
#include <iostream.h>

int main()
{
   long a = 10000;
   long b = 0, d = 0, e = 0, g = 0;
   long c = 2800;
   long f[2801];

   for(;b-c;)
   f[b++] = a/5;

   for(;d=0,g=c*2;c-=14,printf("%.4d",e+d/a),e=d%a)
   for(b=c;d+=f[b]*a,f[b]=d%--g,d/=g--,--b;d*=b);

   return 0;
}
```

The output is

```
3141592653589793238462643383279502884197169399375105820
9749445923078164062862089986280348253421170679821480865
1328230664709384460955058223172535940812848111745028410
2701938521105559644622948954930381964428810975665933446
1284756482337867831652712019091456485669234603486104543
2664821339360726024914127372458700660631558817488152092
0962829254091715364367892590360011330530548820466521384
1469519415116094330572703657595919530921861173819326117
9310511854807446237996274956735188575272489122793818301
1949129833673362440656643086021394946395224737190702179
8609437027705392171762931767523846748184676694051320005
6812714526356082778577134275778960917363717872146844090
1224953430146549585371050792279689258923542019956112129
0219608640344181598136297747713099605187072113499999983
729780499510597317328160963185
```

Chapter 13

Exception Handling

13.1 Introduction

It is a good idea to seperate a program into distinct subsystems that either complete successfully or fail. Thus, a limited form of local error checking should be implemented throughout the system in such a way that the overheads in development time, execution time, memory requirements and overall complexity do not become too high. A fault-tolerant system should be designed hierarchically, with each level coping with as many errors as it can (without making the system too contorted) and leaving the remaining errors to be handled by higher levels of the system. One generally has however the problem that the error cannot be handled locally. The user of the class library knows what to do when an error occurs, but he does not know how to detect an error in the classes. The author of the class library knows how to detect errors, but he does not know how the error should be handled. We thus need a global error communication system. Typically this would be one of the following:

- Error state = return value of function/class method.
- Classes with error state variables.
- Exception handling.

13.2 Error States

One can implement error states via class variables. Typically each object has an error flag (error state variable). If an error occurs the error state variable set to a value indicating error type. For example, the standard IOstream library of C++ has integer state variable `state`. The different bits of this state variable indicate different error states. The error state is queried via the class methods `bad()`, `eof()`, `fail()`. The user can query the error state and implement his own error handling mechanism.

13.3 Exception Handling

A more sophisticated method facilitating the split in resposibilities between error detection and error handling is provided by C++'s support for exception handling. Exceptions handling mechanisms are non-local. They provide a means of communicating errors between library classes/functions and the programs making use of these libraries.

When using exception handling, the library function that detects an error throws an exception in the hope that the calling function is able to catch and handle the exception. For example for a `Rational` class developed we would have a `DivideByZero` exception if the denominator of a rational object becomes zero

```
class Rational
{
  public:
    Rational(const int Numer,const int Denom) ;
    ...
    int numerator();
    int denominator();
    ...
    Rational operator + (const Rational& r)
    ...
    class DivideByZero {};      // exception class
  private:
    int numer, denom;
}
```

Objects of class `DivideByZero` are to be used for exceptions in the following way

```
Rational::Rational(const int Numer,const int Denom) :
                    numer(Numer), denom(Denom)
{
  if(Denom == 0)
    throw DivideByZero();
}
```

A function that wants to detect `DivideByZero` exceptions encloses the code in which it wants to detect the exception in a `try`-block:

```
void main()
{
  ...
  try
  {
  int a, b;
```

```
    ...
   calc(a,b);
    ...
   Rational r1(a,b);
   }
   catch(Rational::DivideByZero)
   {
      ...        // error handling code
   }
   ...
}
```

The catch-block is the actual exception handler. It has to be placed directly after a try block. Note that the catch-block specifies the type of exception it handles. We might also want to throw range errors (when either the denominator or the numerator exceeds the range covered by long int). Such an out-of-range exception could typically occur during addition of two rational numbers. In this case the body of the operator + would throw the exception. In order to distinguish between DivideByZero exceptions and OutOfRange exceptions we would do the following

```
class Rational
{
  public:
    Rational(const int Numer,const int Denom) ;
    ...
    int numerator();
    int denominator();
    ...
    Rational operator + (const Rational& r)
    ...
    class DivideByZero {};      // exception class for /0 error
    class OutOfRange   {};      // exception class for range error
  private:
    int numer, denom;
}
```

The calling program could catch these exceptions in the following way

```
void main()
{
  ...
  try
  {
    int a, b, c, d;
    ...
    calc1(a,b);
```

```
    calc2(c,d);
    ...
    Rational r1(a,b), r2(c,d);
    Rational r3 = r1 + r2;
  }
  catch(Rational::DivideByZero)
  {
    ...          // error handler for /0 errors
  }
  catch(Rational::OutOfRange)
  {
    ...          // error handler for range errors
  }
  ...
}
```

In order to allow for various levels of error specialization/abstraction it is usually a good idea to develop a hierarchical specialization (inheritence) structure of exception classes.

The system classes can now throw an exception as soon as they detect an error. For example, the subscription operator of the `Vector` class can detect a range error and if he does, he throws the relevant exception class

```
template <class T>
inline T& Vector<T>::operator[] (const int i) const
{
  #if EXCEPTIONHANDLING
    if((i<0) || (i>=_length))
      throw Range("Vector::operator[]:");
  #endif
  return _v[i];
}
```

Note that we use a compiler directive which allows us to switch exception-handling on or off. Typically, during the software development, testing and maintenance stages exception-handling is switched on via the precompiler macro

```
#define EXCEPTIONHANDLING 1
```

and once it has been established that the system is stable (and that it conforms to its specifications), exception handling is switched off via

```
#define EXCEPTIONHANDLING 0
```

The resulting executable code will typically run marginally faster (sometimes significantly faster) since it no longer has to run the error-checking code. The user can catch errors via the relevant `try` block

```
try
{
  Vector<int> v5(123);
  ...
  int i;
  for(i=0;;i++)
    cout << v5[i] << " ";
  ...
}
catch(Range rangeError)
{
cout << "*** RANGE ERROR ***: " << rangeError.source << endl;
}
```

Note that the hierarchical exception classes allow for various levels of error specialization. For example, the user could catch all errors via

```
catch(Exception error)
{
  cout << "*** ERROR ***: " << error.source << endl;
}
```

or he could catch all `MathErrors` (e.g. divide-by-zero and math overflow errors) by specifying that he wants to catch all exception classes derived from `MathError` (including the base class `MathError` itself). This is done in the following code excerpt:

```
catch(MathError mathError)
{
cout << "*** ERROR ***: " << mathError.source << endl;
}
```

Of course, within the catch block the user can include any error correction / error handling code he may wish to include.

In the following we implement exception handlong for a `Rational` class which describes rational numbers and their operations.

```
// TRATIONL.CPP
#include <iostream.h>
#include "Rational.h"

void main()
{
  Rational<int> r1(3,9);
  cout << "r1 = " << r1 << endl;
  Rational<int> r2(13,2);
  cout << "r2 = " << r2 << endl;
  Rational<int> r3 = r1 + r2;     // default constructor
  cout << "r3 = r1+r2 = " << r3 << endl;
  cout << "r1*r2 = " << r1*r2 << endl;
  cout << "r1/r2 = " << r1/r2 << endl;
  cout << "-r1 = " << -r1 << endl;
  cout << "r1>r2  -> " << (r1>r2)  << endl;
  cout << "r1<=r2 -> " << (r1<=r2) << endl;
  double x = r1;
  cout.precision(15);
  cout << "x = " << x << endl;
  cout << "r1 + 2 = " << r1 + 2.1 << endl;
  cout << "2 + r1 = " << (int)2 + r1 << endl;
  cout << "r1++ = " << r1++ << endl;
  cout << "r1 = " << r1 << endl;
  cout << "++r1 = " << ++r1 << endl;
  cout << "r1^5  = " << r1.power((int)3) << endl;
  cout << "r1^-0.5 = " << r1.power((int)-0.5) << endl;
  cout << "r1 < 2.3 = " << (r1 < 2.3) << endl;
  Rational<int> r4 = -17;
  cout << "fabs(" << r4 << ") = " << fabs(r4) << endl;
  cout << "Type in rational number r=n/m: n m = ";

  try
  {
  cin >> r1;
  cout << "Read in rational number, r1 = " << r1 << endl;
  }
  catch(DivideByZero error)
  {
  cout << "*** ERROR ***: " << error.source << endl;
  }
 }  // end main
```

Next we provide the header file rational.h.

```
// RATIONAL.H

#ifndef __RATIONAL_H
#define __RATIONAL_H

#include <math.h>
#include <iostream.h>
#include "errhandl.h"

template <class T>
class Rational
{
public:
Rational();
Rational(const T& numer,const T& denom);
Rational(const Rational<T>& r);
Rational(const T& k);

inline T numerator() const;
inline T denominator() const;
Rational<T> power(const long) const;
double power(const double) const;
Rational<T>& operator = (const Rational<T> r);
Rational<T>& operator = (const T& k);
Rational<T> operator + (const Rational<T>& r) const;
Rational<T> operator - (const Rational<T>& r) const;
Rational<T> operator * (const Rational<T>& r) const;
Rational<T> operator / (const Rational<T>& r) const;
Rational<T>& operator += (const Rational<T>& r);
Rational<T>& operator -= (const Rational<T>& r);
Rational<T>& operator *= (const Rational<T>& r);
Rational<T>& operator /= (const Rational<T>& r);
friend Rational<T> operator + (const T& k,const Rational<T>& r);
Rational<T> operator +  (const T& k) const;
Rational<T> operator -  (const T& k) const;
Rational<T> operator *  (const T& k) const;
Rational<T> operator /  (const T& k) const;
double operator + (const double& x) const;
double operator - (const double& x) const;
double operator * (const double& x) const;
double operator / (const double& x) const;
Rational<T>& operator += (const T& k);
Rational<T>& operator -= (const T& k);
```

```
Rational<T>& operator *= (const T& k);
Rational<T>& operator /= (const T& k);
int operator == (const Rational<T>& r) const;
int operator != (const Rational<T>& r) const;
int operator <  (const Rational<T>& r) const;
int operator >  (const Rational<T>& r) const;
int operator <= (const Rational<T>& r) const;
int operator >= (const Rational<T>& r) const;
int operator == (const double& x) const;
int operator != (const double& x) const;
int operator <  (const double& x) const;
int operator >  (const double& x) const;
int operator <= (const double& x) const;
int operator >= (const double& x) const;
int operator == (const T& k) const;
int operator != (const T& k) const;
int operator <  (const T& k) const;
int operator >  (const T& k) const;
int operator <= (const T& k) const;
int operator >= (const T& k) const;
Rational<T> operator + () const;
Rational<T> operator - () const;
Rational<T>& operator++ ();
Rational<T>& operator-- ();
Rational<T>  operator++ (int);
Rational<T>  operator-- (int); // decrement
operator double () const;      // type conversion to double
friend Rational<T> operator + (const T& k,const Rational<T>& r);
friend Rational<T> operator - (const T& k,const Rational<T>& r);
friend Rational<T> operator * (const T& k,const Rational<T>& r);
friend Rational<T> operator / (const T& k,const Rational<T>& r);
friend Rational<T> fabs(const Rational<T>& r);
friend ostream& operator << (ostream& os,const Rational<T>& r);
friend istream& operator >> (istream& is,Rational<T>& r);

private:
   T _numer, _denom;
   T gcd(T a,T b) const;
};

template<class T> Rational<T>::Rational(): _numer(0), _denom(1) {};
template<class T> Rational<T>::Rational(const T& numer,const T& denom)
                : _numer(numer), _denom(denom)
{
#if EXCEPTIONHANDLING
```

```
   if(denom == 0)
     throw DivideByZero("Rational(numer,denom) => denom=0");
#endif
   if(_denom < 0)
     {_numer = -_numer;  _denom = -_denom;}
   T common = gcd(_numer, _denom);
   if(common > 1)
     {_numer /= common;  _denom /= common;}
};

template <class T>
Rational<T>::Rational(const Rational<T>& r)
  : _numer(r._numer), _denom(r._denom) {}

template <class T>
Rational<T>::Rational(const T& k)
  : _numer(k), _denom(1) {}

// Query functions
template <class T>
inline T Rational<T>::numerator() const
  { return _numer; }

template <class T>
inline T Rational<T>::denominator() const
  { return _denom; }

// Other constant member functions
template <class T>
Rational<T> Rational<T>::power(const int n) const
{
if(n==0)
return Rational<T>(1,1);
else
{
Rational<T> result(*this);
int i;
for(i=2; i<=labs(n); i++)
result *= *this;
if(n < 0)
{
T dummy = result._denom;
result._denom = result._numer;
result._numer = dummy;
}
```

```
return result;
}
}

template <class T>
inline double Rational<T>::power(const double y) const
{
return pow(*this,y);
}

// Assignment operators
template <class T>
inline Rational<T>& Rational<T>::operator = (const Rational<T> r)
{
_numer = r._numer; _denom = r._denom;
return *this;
}

template <class T>
inline Rational<T>& Rational<T>::operator = (const T& k)
{
_numer = k; _denom = (T)1;
return *this;
}

// Arithmetic operators
template <class T>
inline Rational<T> Rational<T>::operator + (const Rational<T>& r) const
{
return Rational<T>(_numer*r._denom + r._numer*_denom,_denom*r._denom);
}

template <class T>
inline Rational<T> Rational<T>::operator + (const T& k) const
{
return Rational<T>(_numer+k*_denom,_denom);
}

template <class T>
inline Rational<T> Rational<T>::operator - (const Rational<T>& r) const
{
return *this + (-r);
}

template <class T>
```

```
inline Rational<T> Rational<T>::operator - (const T& k) const
{
  return Rational<T>(_numer-k*_denom,_denom);
}

template <class T>
inline Rational<T> Rational<T>::operator * (const Rational<T>& r) const
{ return Rational<T>(_numer*r._numer, _denom*r._denom); }

template <class T>
inline Rational<T> Rational<T>::operator * (const T& k) const
{
  return Rational<T>(_numer*k,_denom);
}

template <class T>
inline Rational<T> Rational<T>::operator / (const Rational<T>& r) const
{ return Rational<T>(_numer*r._denom, _denom*r._numer); }

template <class T>
inline Rational<T>& Rational<T>::operator += (const Rational<T>& r)
{ *this = *this + r; return *this; }

template <class T>
inline Rational<T>& Rational<T>::operator -= (const Rational<T>& r)
{ *this = *this - r; return *this; }

template <class T>
inline Rational<T>& Rational<T>::operator *= (const Rational<T>& r)
{ *this = *this * r; return *this; }

template <class T>
inline Rational<T>& Rational<T>::operator /= (const Rational<T>& r)
{ *this = *this / r; return *this; }

template <class T>
inline Rational<T>& Rational<T>::operator += (const T& k)
{ *this = *this + k; return *this; }

template <class T>
inline Rational<T>& Rational<T>::operator -= (const T& k)
{ *this = *this - k; return *this; }

template <class T>
inline Rational<T>& Rational<T>::operator *= (const T& k)
```

```
{ *this = *this * k; return *this; }

template <class T>
inline Rational<T>& Rational<T>::operator /= (const T& k)
{ *this = *this / k; return *this; }

template <class T>
inline double Rational<T>::operator + (const double& x) const
{ return (double)*this+x; }

template <class T>
inline double Rational<T>::operator-  (const double& x) const
{
return (double)*this-x;
}

template <class T>
inline double Rational<T>::operator * (const double& x) const
{
return (double)*this * x;
}

template <class T>
inline double Rational<T>::operator / (const double& x) const
{
return (double) *this/x;
}

// Unary operators
template <class T> inline Rational<T>& Rational<T>::operator++ ()
{
*this += 1;
return *this;
}

template <class T> inline Rational<T> Rational<T>::operator++ (int)
{
Rational<T> old(*this); *this+=1;  return old;
}

template <class T> inline Rational<T>& Rational<T>::operator-- ()
{ *this-=1;  return *this; }

template <class T> inline Rational<T> Rational<T>::operator-- (int)
{ Rational<T> old(*this); *this-=1;  return old; }
```

```
template <class T> inline Rational<T> Rational<T>::operator- () const
{
  Rational<T> r(*this);
  r._numer = -r._numer;
  return r;
}

template <class T> inline Rational<T> Rational<T>::operator+ () const
{
return *this;
}

// Relational operators
template <class T>
inline int Rational<T>::operator == (const Rational<T>& r) const
{ return ((_numer == r._numer) && (_denom == r._denom)); }

template <class T>
inline int Rational<T>::operator != (const Rational<T>& r) const
{ return (!(*this == r)); }

template <class T>
inline int Rational<T>::operator > (const Rational<T>& r) const
{ return ((double)*this > (double)r); }

template <class T>
inline int Rational<T>::operator < (const Rational<T>& r) const
{ return ((double)*this < (double)r); }

template <class T>
inline int Rational<T>::operator >= (const Rational<T>& r) const
{ return ((double)*this >= (double)r); }

template <class T>
inline int Rational<T>::operator <= (const Rational<T>& r) const
{ return ((double)*this <= (double)r); }

template <class T>
inline int Rational<T>::operator == (const T& k) const
{ return (_numer == k); }

template <class T>
inline int Rational<T>::operator != (const T& k) const
{ return (_numer != k); }
```

```
template <class T>
inline int Rational<T>::operator> (const T& k) const
{
return ((double)*this > k);
}

template <class T> inline int Rational<T>::operator<(const T& k) const
{
return ((double)*this < k);
}

template <class T>
inline int Rational<T>::operator >= (const T& k) const
{
return ((double)*this >= k);
}

template <class T>
inline int Rational<T>::operator <= (const T& k) const
{ return ((double)*this <= k); }

template <class T>
inline int Rational<T>::operator == (const double& x) const
{
return ((double)*this == x);
}

template <class T>
inline int Rational<T>::operator != (const double& x) const
{ return ((double)*this != x); }

template <class T>
inline int Rational<T>::operator > (const double& x) const
{ return ((double)*this > x); }

template <class T>
inline int Rational<T>::operator < (const double& x) const
{ return ((double)*this < x); }

template <class T>
inline int Rational<T>::operator >= (const double& x) const
{ return ((double)*this >= x); }

template <class T>
```

```
inline int Rational<T>::operator <= (const double& x) const
{ return ((double)*this <= x); }

// Type conversion
template <class T> inline Rational<T>::operator double () const
{ return (double)_numer/(double)_denom; }

// Private methods
template <class T> T Rational<T>::gcd(T a,T b) const
{
  if(a < 0) a = -a;
  if(b < 0) b = -b;

  while(b > 0)
  {
  T m = a%b;
  a = b;
  b = m;
  }

  return a;
}

// Friends of Rational
template <class T>
ostream& operator<< (ostream& os,const Rational<T>& r)
{
T non_frac = r._numer/r._denom;
T frac_numer = r._numer - non_frac*r._denom;
if(non_frac != 0) os << non_frac;
if(frac_numer != 0)
if(frac_numer > 0)
os << " " << frac_numer << "/" << r._denom;
else
os << " " << -frac_numer << "/" << r._denom;
return os;
}

template <class T> istream& operator >> (istream& is,Rational<T>& r)
{
T numer, denom;
is >> numer >> denom;
r = Rational<T>(numer,denom);
return is;
}
```

```
template <class T>
Rational<T> operator + (const T& k,const Rational<T>& r)
{ return r+k; }

template <class T>
Rational<T> operator- (const T& k,const Rational<T>& r)
{ return (-r)+k; }

template <class T>
Rational<T> operator* (const T& k,const Rational<T>& r)
{ return r*k; }

template <class T>
Rational<T> operator / (const T& k,const Rational<T>& r)
{ return Rational<T>(r._denom*k,r._numer); }

template <class T> Rational<T> fabs(const Rational<T>& r)
{ return Rational<T>(fabs(r._numer),r._denom); }
#endif
```

```
#ifndef __ERRHANDL_H
#define __ERRHANDL_H
#define EXCEPTIONHANDLING 1

// BASE CLASS FOR ALL EXCEPTION HANDLING
class Exception
{
    public:
      char* source;
      Exception(): source(NULL) {}
      Exception(char* src): source(src) {}
};

// BASE CLASS FOR ALL MATH ERRORS
class MathError: public Exception
{
    public:
      MathError() {}
      MathError(char* src): Exception(src) {}
};

// NOW ALL THE SPECIAL INSTANCES OF MATH ERRORS
class DivideByZero: public MathError
{
   public:
     DivideByZero() {}
     DivideByZero(char* src) : MathError(src) {}
};

class Overflow: public MathError
{
   public:
      Overflow() {}
      Overflow(char* src) : MathError(src) {}
};

class IllegalOperation: public MathError
{
   public:
     IllegalOperation() {}
     IllegalOperation(char* src) : MathError(src) {}
};

// BASE CLASS FOR ALL MEMORY ERRORS
class MemoryError: public Exception
```

```
{
  public:
    MemoryError() {}
    MemoryError(char* src): Exception(src) {}
};

// NOW ALL THE SPECIAL INSTANCES OF MEMORY ERRORS
class OutOfMemory: public MemoryError
{
  public:
    OutOfMemory() {};
    OutOfMemory(char* src) : MemoryError(src) {}
};

class Range: public MemoryError
{
   public:
    Range() {}
    Range(char* src) : MemoryError(src) {}
};

// SOME OTHER EXCEPTIONS
class IllegalArguments: public Exception
{
  public:
    IllegalArguments() {}
    IllegalArguments(char* src): Exception(src) {}
};

class IllegalCall: public Exception
{
  public:
   IllegalCall() {}
   IllegalCall(char* src): Exception(src) {}
};
#endif
```

Chapter 14

The Standard Template Library

14.1 Introduction

When templates were introduced in the C++ language some years ago, few C++ programmers were aware of the influence this concept would have on the standard library. The *Standard Template Library* (STL) is one of the largest and most significant changes to have come out of this process. The intent of the STL is to provide a set of container classes that are both efficient and functional. The presence of such a library will simplify the creation of complex programs, and because the library is standard the resulting programs ultimately will have a high degree of portability. This chapter is about this new standard version of STL. The use of STL is likely to make software more reliable, more portable and more general and to reduce the cost of producing it.

One of the most interesting aspects of the STL is the way it radically departs in structure from almost all earlier libraries. Since C++ is an object-oriented language, these earlier libraries, have tended to rely heavily on object-oriented techniques, such as inheritance. The STL uses almost no inheritance. In many ways the STL is almost anti-object-oriented. To see this non-object-oriented perspective, consider that object-oriented programming holds encapsulation as a primary ideal. A well-designed object will try to encapsulate all the state and behavior necessary to perform the task for which it is designed, and at the same time, hide as many of the internal implementation details as possible. In almost all previous object-oriented container class libraries this philosophical approach was manifested by collection classes with exceedingly rich functionality and consequently, with large interfaces and complex implementations. The designers of STL moved in an entirely different direction. Each component is designed to operate in conjunction with a rich collection of generic algorithms. These generic algorithms are independent of the containers and can therefore operate with many different containers types. By separating the functionality of the generic algorithms from the container classes themselves, the STL realizes a great savings in size, in both the library and the generated code.

Instead of duplication of algorithms in each of the dozen or so different container classes, a single definition of a library function can be used with any container. Furthermore, the definition of these functions is so general that they can be used with ordinary C-style arrays and pointers as well as with other data types. The boolean data type `bool` is also implemented in the Standard Template Library.

At the core of the Standard Template Library are three foundational items:

1) containers

2) algorithms

3) iterators.

These items work in conjunction with one another. Containers are objects that hold other objects. Algorithms act on the contents of containers. Iterators are objects that are pointers. They provide us with the ability to cycle through the contents of a container in the same way that we apply pointers to cycle through an array.

The classes provided by the standard template library are:

```
algorithm
deque
functional
iterator
list
map
memory
numeric
queue
set
stack
utility
vector
```

14.2 The Namespace Concept

Namespaces are a recent addition to C++. A namespace creates a declarative region in which various program elements can be placed. Elements declared in one namespace are separated from elements declared in another. The line `using namespace std;` tells the compiler to use the std namespace.

```
// namesp.cpp

#include <iostream>
using namespace std;

namespace A
{
  int i = 10;
}

namespace B
{
  int i = 5;
}

void fA()
{
   using namespace A;
   cout << "In fA: " << A::i << " " << B::i << " " << i << endl;
   // => 10 5 10
}

void fB()
{
  using namespace B;
  cout << "In fB: " << A::i << " " << B::i << " " << i << endl;
  // => 10 5 5
}

int main()
{
   fA();  fB();
   cout << A::i << " " << B::i << endl; // => 10 5
   using A::i;
   cout << i << endl; // => 10

   return 0;
}
```

14.3 The Vector Class

The vector class is a container class. The vector class supports a dynamic array. This is an array that can grow as needed. We can use the standard array subscript notation [] to access its elements.

The constructors and methods (member functions) in the class `vector` are summarized in the following.

```
Constructors
vector<T> v;                   default construction
vector<T> v(int);              initialized with explicit size
vector<T> v(int,T);            size and initial value
vector<T> v(aVector);          copy constructor

Element Access
v[i]                           subscript access, can be assignment target
v.front()                      first value in collection
v.back()                       last value in collection

Insertion
v.push_back(T)                 push element on to back of vector
v.insert(iterator,value)       insert new element after iterator
v.swap(vector<T>)              swap values with another vector

Removal
v.pop_back()                   pop element from back of vector
v.erase(iterator)              remove single element
v.erase(iterator,iterator)     remove range of values

Size
v.capacity()                   maximum number of elements buffer can hold
v.size()                       number of elements currently held
v.resize(unsigned,T)           change to size, padding with value
v.reserve(unsigned)            set physical buffer size
v.empty()                      true if vector is empty

Iterators
vector<T>::iterator itr        declare a new iterator
v.begin ()                     starting iterator
v.end ()                       ending iterator
```

```
// Mvector.cpp

#include <iostream>
#include <vector>
using namespace std;

int main()
{
   vector<char> w(4);
   w[0] = 'X';   w[1] = 'Y';   w[2] = '+';   w[3] = '-';
   cout << w[2] << endl;  // => +

   int j;
   for(j=0; j<4; j++)
   cout << "w[" << j << "] = " << w[j] << endl;

   vector<double> v;
   double x;
   cout << "enter double number, followed by 0.0:\n";
   while(cin >> x, x != 0.0)
   v.push_back(x);     // adds the double value x at the end
                       // of the vector v
   vector<double>::iterator i;
   for(i=v.begin(); i != v.end(); ++i)
   cout << *i << endl; // dereferencing

   bool b = v.empty;
   cout << "b = " << b << endl;

   // insert element in the vector
   double y;
   cout << "enter value to be inserted: ";
   cin >> y;
   vector<double>::iterator p = v.begin();
   p += 2;
   v.insert(p,4,y);  // point to 3rd element, insert 4 elements

   for(i=v.begin(); i != v.end(); ++i)
   cout << *i << endl;

   int length = v.size();
   cout << "length = " << length << endl;

   return 0;
}
```

We read a textfile line by line into a vector. We read the file, line by line, until `eof` is encountered. The only catch is that with the istream's operator `>>` string can't be used. The function only reads a word, and not a whole line. The global

```
getline(istream&, string&)
```

is the solution. Let the text file (ASCII file) be `william.txt` with the contents

```
to be or
not to be
```

After compiling and linking to obtain the execute file at the command line we enter:

```
readin1 william.txt
```

```
// readin1.cpp

#include <iostream>
#include <fstream>
#include <string>
#include <vector>

using namespace std;

int main(int argc, char *argv[])
{
   ifstream ifs(argv[1]);
   vector<string> v;
   string s;

   while(getline(ifs,s))
   {
     v.push_back(s);
     cout << s <<endl;
   }

   return 0;
}
```

```
// bubblest.cpp

#include <algorithm>
#include <vector>
#include <stdlib.h>
#include <iostream.h>

using namespace std;

template <class RandomAccessIterator>
void bubbleSort(RandomAccessIterator first,RandomAccessIterator last)
{
   int exchanged;
   do
   {
   exchanged = 0; last--;
   for(RandomAccessIterator iter=first; iter < last; iter++)
   {
   if(*iter > *(iter+1))
   {
   iter_swap(iter,iter+1);
   exchanged++;
   }
   }
   } while(exchanged);
}

void main(void)
{
   const int nData = 6;
   vector<unsigned long> data(nData);

   for(int n=0; n < nData; n++)
   {
   data[n] = rand();
   cout << data[n] << " ";
   }
   cout << endl;

   bubbleSort(data.begin(),data.end());

   // after sorting
   for(n=0; n<nData; n++)
   cout << data[n] << " ";
}
```

14.4 The List Class

The list class is also a container class. A linked list is the data structure we choose when the number of elements in a collection cannot be bounded, or varies widely during the course of execution. Like the vector class, the linked list class maintains values of uniform type. Lists are not indexed. The following table gives the constructors and the member functions of the class.

Constructors and Assignment	
list<T> v;	default constructor
list<T> v(aList);	copy constructor
l = aList	assignment
l.swap (aList)	swap values with another list
Element Access	
l.front()	first element in list
l.back()	last element in list
Insertion and Removal	
l.push_front(value)	add value to front of list
l.push_back(value)	add value to end of list
l.insert(iterator, value)	insert value at specified location
l.pop_front()	remove value from front of list
l.pop_back()	remove value from end of list
l.erase(iterator)	remove referenced element
l.erase(iterator,iterator)	remove range of elements
l.remove(value)	remove all occurrences of value
l.remove_if(predicate)	removal all values that match condition
Size	
l.empty()	true if collection is empty
l.size()	return number of elements in collection
Iterators	
list<T>::iterator itr	declare a new iterator
l.begin()	starting iterator
l.end()	ending iterator
l.rbegin()	starting backwards moving iterator
l.rend()	ending backwards moving iterator
Miscellenous	
l.reverse()	reverse order of elements
l.sort()	place elements into ascending order
l.sort(comparison)	order elements using comparison function
l.merge(list)	merge with another ordered list

The following program gives an application of the list class.

```
// mlist.cpp

#include <iostream>
#include <list>
#include <string>
using namespace std;

void main(void)
{
   list<char> lst;
   int i;
   for(i=0; i<10; i++)
   lst.push_back('X'+i);
   cout << "size of list = " << lst.size() << endl;
   cout << "contents of list: ";
   list<char>::iterator p = lst.begin();

   while(p != lst.end())  {  cout << *p << " ";  p++;  }
   lst.push_back('A');
   cout << endl;

   lst.sort();
   cout << "sorted contents:\n";
   p = lst.begin();
   while(p != lst.end())  {  cout << *p << " ";  p++;  }
   cout << endl;

   list<string> str;
   str.push_back("Good");
   str.push_back(" ");
   str.push_back("Night");
   str.push_back(" ");
   str.push_back("Egoli");
   list<string>::iterator q = str.begin();

   while(q != str.end())  {  cout << *q << " ";  q++;  }
}
```

14.5 The Stack Class

A stack is a collection of data items organized in a linear sequence, together with the following five operations:

1) *CreateStack* brings a stack into existence

2) *MakeStackEmpty* deletes all items, if any, from the stack

3) *Push* adds an item at one end, called the top, of the stack

4) *Pop* removes the item at the top of a stack and makes it available for use

5) *StackIsEmpty* tests whether a stack contains any items.

The items in a stack might be integers, real numbers, characters, or abstract data types.

The following list give the operations provided by the standard template library for the stack class.

```
Insertion and Removal
s.push(value)      push value on front of stack
s.top()            access value at front of stack
s.pop()            remove value from front of stack

Size
s.size()           number of elements in collection
s.empty()          true if collection is empty

Comparisons
==
!=
<
<=
>
>=
```

The following program shows a simple application of the stack class.

```
// stack1.cpp

#include <iostream>
#include <vector>
#include <stack>
using namespace std;

int main()
{
   stack<char,vector<char> > S, T, U;
   cout << "&S = " << &S << endl;

   S.push('X');
   S.push('Y');
   S.push('Z');
   cout << "characters pushed onto S: X Y Z\n";

   T = S;
   cout << "after T = S; we have ";
   cout << (S == T ? "S == T" : "S != T") << endl;

   U.push('X');
   U.push('W');
   cout << "pushed onto U: X W\n";
   cout << "output now ";
   cout << (S < U ? "S < U" : "S >= U") << endl;

   bool b = (T == S);
   cout << "b = " << b << endl;

   return 0;
}
```

14.6 The Queue Class

A queue is a collection of data items organized in a linear sequence, together with the following five operations:

1) *CreateQueue* brings a queue into existence

2) *MakeQueueEmpty* deletes all items, if any, from the queue

3) *EnQueue* adds an item at one end, called the rear, of the queue

4) *DeQueue* removes the item from the other end, called the front, of the queue, and makes it available for use

5) *QueueIsEmpty* tests whether a queue is empty

The following list give the operations provided by the standard template library for the stack class.

```
Insertion and Removal
q.push(value)       push value on back of queue
q.front()           access value at front of queue
q.back()            access value at back of queue
q.pop()             remove value from front of queue

Size
q.size()            number of elements in collection
q.empty()           true if collection is empty
```

The following program shows a simple application of the queue class.

```
// queue1.cpp

#include <iostream>
#include <list>
#include <queue>

using namespace std;

int main()
{
   queue<double,list<double> > Q;
   cout << "&Q = " << &Q << endl;

   Q.push(3.14);
   Q.push(4.5);
   Q.push(6.9);
   cout << "after pushing 3.14, 4.5, 6.9:\n";
   cout << "Q.front() = " << Q.front() << endl;
   cout << "Q.back() = " << Q.back() << endl;

   Q.pop();
   cout << "after Q.pop():\n";
   cout << "Q.front() = " << Q.front() << endl;

   return 0;
}
```

14.7 The Deque Class

A deque (double-ended queue) is an abstract data type in which insertions and deletions can be made at either the front or the rear. Thus it combines features of both stacks and queues. In practice, the deque has two variations. The first is the input-restricted deque, where insertions are restricted to one end only, and the second is the output restricted deque, where deletions are restricted to a single end.

The following list give the operations provided by the standard template library for the stack class.

```
Constructors and Assignment
deque<T> d;                   default constructor
deque<T> d(int);              construct with initial size
deque<T> d(int,value);        construct with initial size and initial value
deque<T> d(aDeque);           copy constructor
d = aDeque;                   assignment of deque from another deque
d.swap(aDeque);               swap contents with another deque

Element Access and Insertion
d[i]                          subscript access, can be assignment target
d.front()                     first value i collection
d.back()                      final value in collection
d.insert(iterator,value)      insert value before iterator
d.push_front(value)           insert value at front of container
d.push_back(value)            insert value at back of container

Removal
d.pop_front()                 remove element from front of vector
d.pop_back()                  remove element from back of vector
d.erase(iterator)             remove single element
d.erase(iterator,iterator)    remove range of elements

Size
d.size()                      number of elements currently held
d.empty()                     true if vector is empty

Iterators
deque<T>::iterator itr        declare a new iterator
d.begin()                     starting iterator
d.end()                       stopping iterator
d.rbegin()                    starting iterator for reverse access
d.rend()                      stopping iterator for reverse access
```

The following small program shows an application of the deque class.

```
// deque1.cpp

#include <iostream>
#include <deque>
using namespace std;

int main()
{
   deque<char> D(10,'A');

   deque<char>::iterator i;
   for(i=D.begin(); i != D.end(); ++i)
   cout << *i << " ";
   cout << endl;

   deque<char> E(D);  // copy constructor

   E.push_front('Z');
   E.push_back('Y');

   for(i=E.begin(); i != E.end(); ++i)
   cout << *i << " ";
   cout << endl;

   cout << E.empty() << endl;

   cout << E.size() << endl;

   char c = E[0];
   cout << "c = " << c << endl;

   c = E[5];
   cout << "c = " << c << endl;


   return 0;
}
```

14.8 The Bit Set Class

The bit set class in the standard template library implements the bitwise operations.

```
Constructors
bitset<N> s              construct bitset for N bits
bitset<N> s(aBitSet)     copy constructor

Bit level operations
s.flip()                 flip all bits
s.flip(i)                flip position i
s.reset(0)               set all bits to false
s.reset(i)               set bit position i to false
s.set()                  set all bits to true
s.set(i)                 set bit position i to true
s.test(i)                test if bit position i is true

Operations on entire collection
s.any()                  return true if any bit is true
s.none()                 return true if all bits are false
s.count()                return number of true bits

Combination with other bitsets
s1 & s2                  bitwise AND
s1 | s2                  bitwise inclusive OR
s1 ^ s2                  bitwise exclusive OR
s == s2                  return true if two sets are the same

other operations
s << n                   shift set left by one
s >> n                   shift set right by one
s.to_string()            return string representation of set
```

The following small program shows an application of the Bit Set class.

```
// bitset1.cpp

#include <iostream>
#include <bitset>
using namespace std;

int main()
{
   const unsigned long n = 32;
   bitset<n> s;
   cout << s.set() << endl;        // set all bits to 1

   cout << s.flip(12) << endl;     // flip position i

   bitset<n> t;
   cout << t.reset() << endl;      // set all bits to false

   t.set(23);
   t.set(27);

   bitset<n> u;
   u = s & t;
   cout << "u = " << u << endl;

   bitset<n> v;
   v = s | t;
   cout << "v = " << v << endl;

   bitset<n> w;
   w = s ^ t;
   cout << "w = " << w << endl;

   bitset<n> z;
   z = w ^ w;
   cout << "z = " << z << endl;

   return 0;
}
```

14.9 The Set Class

The abstract data type set is defined in terms of objects and operations. The objects are just sets in the mathematical sense with the representation unspecified. The `set` and `multiset` data types in the standard template library are both template data structures, where the template argument represents the type of the elements the collection contains. In the `set` class each element of a `set` is identical with its key, and keys are unique. Because of this, two distinct elements of a set cannot be equal. A multiset differs from a set only in that it can contain equal elements. The operations in the set class are given below.

```
Constructors
set<T> s;                    default constructor
multiset<T> m;               default constructor
set<T> s(aSet);              copy constructor
multiset<T> m(aMultiset)     copy constructor
s = aSet                     assignment
s.swap(aSet)                 swap elements with argument set

Insertion and Removal
s.insert(value_type)         insert new element
s.erase(value_type)          remove all matching elements
s.erase(iterator)            remove element specified by iterator
s.erase(iterator,iterator)   remove range of values

Testing for Inclusion
s.empty()                    true if collection is empty
s.size()                     number of elements in collection
s.count(value_type)          count number of occurrences
s.find(value_type)           locate value
s.lower_bound(value_type)    first occurrence of value
s.upper_bound(value_type)    next element after value
s.equal_range(value_type)    lower and upper bound pair

Iterators
set<T>::iterator itr         declare a new iterator
s.begin()                    starting iterator
s.end()                      stopping iterator
s.rbegin()                   starting iterator for reverse access
s.rend()                     stopping iterator for reverse access
```

The following program shows an application of the set class.

```
// set2.cpp

#include <iostream>
#include <set>
using namespace std;

int main()
{
   set<int> s;

   s.insert(23);
   s.insert(45);
   s.insert(-1);
   s.insert(-2);
   s.insert(23);
   s.insert(51);

   cout << s.empty() << endl;   // => 0 (false)

   cout << s.size() << endl;    // => 5

   cout << s.count(23) << endl; // => 1

   set<int>::iterator i = s.begin();

   for(i=s.begin(); i != s.end(); ++i)
   cout << *i << " ";

   cout << endl;

   s.erase(45);

   for(i=s.begin(); i != s.end(); ++i)
   cout << *i << " ";

   cout << endl;

   cout << s.size() << endl; // => 4

   return 0;
}
```

14.10 The Map Class

The map class (also called dictionary or table) is an indexed collection. The index values need not be integers, but can be any ordered data values. Therefore a map is a collection of associations of key value pairs. For maps no two keys can be equal. A multimap differs from a map in that duplicated keys are allowed. The following table gives the operations.

```
Constructors
multimap<T> m;              default constructor
multimap<T> m;              default constructor
map<T> m(aMap)              copy constructor
multimap<T> m(aMultiMap)    copy constructor
m = aMap                    assignment

Insertion and Removal
m[key]                      return reference to value with key
m.insert(value_type)        insert given key value pair
m.erase(key)                erase value with given key
m.erase(iterator)           erase value at given iterator

Testing for Inclusion
m.empty()                   true if collection is empty
m.size()                    return size of collection
m.count(key)                count number of elements with given key
m.find(key)                 locate element with given key
m.lower_bound(key)          first occurrence of key
m.upper_bound(key)          next element after key
m.equal_range(key)          lower and upper bound pair

Iterators
map<T>::iterator itr;       declare new iterator
m.begin()                   starting iterator
m.end()                     ending iterator
m.rbegin()                  backwards moving iterator start
m.rend()                    backwards moving iterator end
```

The following two programs show applications of the map class.

```
// mmap.cpp

#include <iostream>
#include <map>
#include <string>
using namespace std;

int main()
{
   map<char,int> m1;
   int i;

   for(i=0; i<26; i++)
   {
   m1.insert(pair<char,int>('A'+i,65+i)); // ASCII table
   }

   char ch;
   cout << "enter key: ";
   cin >> ch;

   map<char, int>::iterator p1;

   // find value given key
   p1 = m1.find(ch);
   if(p1 != m1.end())
   cout << "ASCII table number: " << p1 -> second << endl;
   else
   cout << "key not in map.\n";

   map<string,double> m2;
   m2.insert(pair<string, double>("Willi",0.0));
   m2.insert(pair<string, double>("Fritz",1.0));
   m2.insert(pair<string, double>("Charles",2.0));

   map<string,double>::iterator p2;

   p2 = m2.find("Fritz");
   cout << p2 -> second << endl;  // => 1
   cout << p2 -> first << endl;   // => Fritz

   return 0;
}
```

```
// mystl.cpp
#include <iostream>
#include <map>
#include <string>
using namespace std;

typedef map<string, string> vocabulary;

void main(void)
{
   string wrd, def;
   vocabulary voc;
   vocabulary::iterator itr;
   cout << "Populating the dictionary" << endl;
   while(true)
   {
      cout << "Enter word: ";
      cout.flush();
      cin >> wrd;
      if((wrd == "Q") || (wrd == "q")) break;
      cout << "Definition: ";
      cout.flush();
      cin >> def;
      voc[wrd] = def;
   }
   if(voc.empty())
   {
      cout << "Empty vocabulary!" << endl;
      exit(0);
   }
   cout << "Vocabulary populated with ";
   cout << voc.size() << " elements." << endl;
   cout << "Looking up words in the dictionary" << endl;
   while(true)
   {
      cout << "Enter word: ";    cin >> wrd;
      if((wrd == "Q") || (wrd == "q")) break;
      itr = voc.find(wrd);
      if(itr != voc.end())
          cout << itr->second << endl;
      else
          cout << "not found!" << endl;
   }
}
```

14.11 The Algorithm Class

Algorithms act on the contents of containers. They include functions for initializing, sorting, searching, and transforming the contents of containers. The functions can also be applied to arrays of basic data types. The functions are:

```
sort, find, copy, merge, reverse, replace
```

The following code shows an application of some of the functions:

```
// algorith1.cpp

#include <iostream>
#include <algorithm>
using namespace std;

int main()
{
   const int n = 5;
   double x[n];
   x[0] = 3.13; x[1] = 4.5; x[2] = 1.2; x[3] = 0.8; x[4] = 0.1;

   sort(x,x+n);
   cout << "array after sorting: \n";
   double* p;
   for(p=x; p != x+n; p++)
   cout << *p << " ";
   cout << endl;

   double y;
   cout << "double value to be searched for: ";
   cin >> y;

   double* q = find(x,x+n,y);
   if(q == x+n)
   cout << "not found\n";
   else
   {
   cout << "found";
   if(q == x)
   cout << " as the first element";
   else
   cout << " after " << *--q;
   }
```

```
  cout << endl;

  reverse(x,x+n);
  double* r;
  cout << "reversed array: " << endl;
  for(r=x; r != x+n; r++)
  cout << *r << " ";
  cout << endl;

  int a[6] = { 2, -1, 6, 7, 15, -8 };
  int b[4] = { 3, 2, -8, 4 };
  int c[10];
  merge(a,a+6,b,b+4,c);
  int i;
  for(i=0; i<10; i++)
  cout << "c[" << i << "] = " << c[i] << endl;

  return 0;
}
```

Bibliography

[1] Ammeraal Leen, *STL for C++ Programmers*, John Wiley, Chichester, 1997

[2] Baase Sara, *Computer Algorithms*, Addison-Wesley, Reading, MA, 1989

[3] Berry John Thomas, *The Waite Group's C++ Programming*, Second Edition, SAMS, Carmel, Indiana, 1992

[4] Budd Timothy, *Classic Data Structures in C++*, Addison-Wesley, Reading, Massachusetts, 1994

[5] Budd Timothy, *Data Structures in C++ Using the Standard Template Library*, Addison-Wesley, Reading, Massachusetts, 1998

[6] Ellis Margaret A. and Stroustrup Bjarne, *The Annotated C++ Reference Manual*, Addison-Wesley, Reading, MA, 1990

[7] Lafore Robert, *Object-Oriented Programming in TURBO C++*, Waite Group Press, Mill Valley, 1991

[8] Lippman Stanley B., *C++ Primer (2nd edition)*, Addison-Wesley, Reading, MA, 1992

[9] Meyers Scott, *Effective C++*, Addison-Wesley, Reading, MA. 1992

[10] Schildt Herbert, *C++ from the Ground Up*, Osborne, Berkely, 1998

[11] Smith Michael A., *Object-Oriented Software in C++*, Chapman and Hall, London, 1993

[12] Steeb Willi-Hans, *Quantum Mechanics Using Computer Algebra*, World Scientific, Singapore, 1994

[13] Stroustrup Bjarne, *The C++ Programming Language*, 2nd Edition, Addison-Wesley, Reading, MA, 1991

[14] Tan Kiat Shi and Willi-Hans Steeb *SymbolicC++: An Introduction to Computer Algebra Using Object-Oriented Programming*, Springer-Verlag, Singapore, 1998

Index

EOF, 321
NULL, 41
argc, 51
argv, 51
assert, 210
atof, 133
atoi, 427
break, 26, 69
bsearch, 171
catch, 486
char, 9
cin, 8, 112
class, 115
const, 9, 67
cout, 8, 291
ctime, 64
delete, 38
difftime, 64
do while, 24
fclose, 308
fopen, 308, 318
fread, 318
friend, 135
fseek, 318
fwrite, 318
getline, 112
goto, 69, 468
if, 22
include, 8
inline, 68
int, 9
iostream, 8
long, 9
lsearch, 172
main, 8
map, 522
new, 38
private, 115, 293
short, 9
sizeof, 30, 37, 150
sprintf, 133
static, 71
strcat, 98
strchr, 414
strcmp, 90
strlen, 92
strncmp, 97
strstr, 88
strtok, 133
struct, 72
switch, 26
this, 121, 288
throw, 486
virtual, 140
void, 8
while, 24

Abstract data types, 189
Address-of-operator, 18
Arithmetic mean, 454
Array, 28
Arrays, 212
ASCII table, 14
Assembly language, 63
Assignment operator, 119
Atom, 263
Autocorrelation function, 477

Basic data types, 9
Binary tree, 269
Bitwise operation, 55
Boyer-Moore algorithm, 104
Bubblesort, 165
Byte, 9

Class constants, 291

Class templates, 151
Comma operator, 329
Comments, 7
Constants, 67
Constructors, 116
Control statements, 21
Copy constructor, 119

Data encryption, 394
Date class, 282, 347
Dereference operator, 18
Destructors, 116
Deviation, 458
Dijkstra algorithm, 359
Dynamic binding, 114

Eliza, 375, 409
Encryption, 394
Error handling, 485
Exception handling, 486, 487

False, 9
Finite set, 276
Finite state machine, 443
Floyd-Warshall algorithm, 356
Function templates, 143
Functions, 8, 143

Generalized list, 263
Geometic mean, 454

Harmonic mean, 454
Hexadecimal notation, 9
Huffman coding, 403

Indirection operator, 18
Information hiding, 2
Inheritance, 2
Inline functions, 67
Integer division, 11
Intrusive list, 258

Key, 163
Keywords, ix

Least square norm, 463
Linear congruential algorithm, 473
Linear correlation coefficient, 475
Linear Programming, 362
Link, 238
Linked list, 237
LISP, 237, 263

Macros, 67
Mean value, 458
Member function, 292
Member-of-operator, 72
Mergesort, 386
Methods, 113
Modulus operator, 11
Month class, 284

Namespaces, 505
Null character, 14, 38

Object, 115
Operands, 11
Operator overloading, xi, 120, 291
Operator precedence, xi
Operator type conversion, 291
Operators, xi

Pass by reference, 48
Pass by value, 48
Permutations, 86
Pointer, 18, 40, 210
Polymorphism, 3
Precedence, 16, 287
Precedence table, xi
Pseudo random number, 473

Quicksort, 165

Recursion, 60
Reference types, 19
Relational operators, 21
Remainder operator, 11
Roman numerals, 427

Searching, 163
Shellsort, 167
Shift operators, 57
Simplex method, 363
Singly linked circular list, 243
Sorting, 163

Stack class, 156
Standard deviation, 458
Standard Template Library, 503
State, 485
Statements, 8
Stream access, 291
Strings, 33, 75
Structure pointer operator, 74

Tableau, 363
Templates, 3
Time average, 477
True, 9

Whitespace, 327